D0848457

Pelagic and Semi-Pelagic Trawling Gear

John Garner

Pelagic and Semi-Pelagic Trawling Gear

British Library CIP Data
Garner, John
Pelagic and semi-pelagic trawling gear.
1. Trawls and trawling
I. Title
639'.22'028 SH344.6.T7

ISBN 0-85238-088-7

Illustrations by the author
Design by Eric Williams
Printed in Great Britain by
The Whitefriars Press Ltd London and Tonbridge

Contents

Acknowledgements

In the course of one's work certain information obviously becomes available, and much of it is catalogued or recorded for future use according to the degree of interest. It is often the case, however, that other people have diligently researched through a particular theory, and later the theory is developed and gradually adopted and perfected for commercial use.

I have endeavoured to contribute to the understanding of the subject of fishing gears, and have at times had to confirm some of my work by reference; consequently, I have often been glad to be able to look up works produced by others occupied in the field of fishing gear technology.

For certain information, which I particularly requested, I am obliged to the Director and G. Thorsteinsson of the Marine Research Institute, Reykjavik. Also, to Mr. Kiyoteru Kimoto, Manager of the Research & Development Department of Hakodate Seimo Sengu Co. Ltd., Nitta Bldg 8.8 Ginza-Nishi, Chuo-Ku, Tokyo, Japan. The information which was supplied is covered by Figures 19, 22, 23, 35, 36 and 37.

JOHN GARNER

List of Illustrations

Fig

Preface

In my first book, *Deep Sea Trawling,* I stated that fish could be caught at certain periods 'off the bottom' with the standard Granton otter trawl gear. That was twenty years ago when towed nets for catching fish that were swimming between the surface and the sea-bed were still in the experimental stages.

Things have improved since then. Electronic equipment in varied forms has come to aid the fishermen. Now available are the most sensitive echo sounders giving to the searching skipper the finest of details. And fischlupes flash and sonars blit with instantaneous outline of large areas of sea lying ahead.

On the mechanical side semi-automatic shooting and hauling machinery is available for handling all types of fishing gear. Powered blocks, automatic winches, safety releases, loading devices and various other assemblies are available and freely used – today, in fact, fish are caught and processed with minimum handling.

Fishing craft themselves are now sophisticated vessels equipped with two-berth cabins, showers, often quite enviable catering facilities, stereo recording units, films and television. Such are the advances of past decades. But the able handling and operation of all this gear and equipment still demands the knowledge, skill and ability of the trained fisherman – he must always be ready to meet and cope with new conditions or unexpected developments.

One outstanding modern factor is over-fishing occasioned by the ever-increasing hunger of rising populations. With the heavy and even over-fishing there arises the threat of eliminating some species, and that threat sounds the clarion call for conservation and preservation of stocks. To that problem more and more consideration must be given. One solution is short-term and quickly rewarding and lies in the periodic closure of known breeding grounds. Other methods are the introduction of international quota systems and multi-lateral agreed landing rates. All these moves are inevitable in common interest and they will be accepted as normal even with their known drawbacks.

Faced with those developments I foresee the growth of 'unitised' operation. By that term I wish to indicate the establishment of a working/catching unit which has achieved a high degree of proficiency and is thus accepted. An example might be the Gulf of Mexico style of fishing for shrimps or prawns where the vessel falls within a certain small range of specification details and with the equipment and use of same more or less standardised. Such a unit can be adopted and adapted for use in the seas of many nations on suitable stocks. And, passing from the ring-net to the adaptable power-block, another example that can be given lies in the range of middle-water purse seines such as those pioneered by Norway and used for fishing the herring stocks of the North Sea.

Because of the industry's changing needs and requirements, stern trawlers built for ground fishing in the particular range of near-water, middle- or deep-sea operations, are now usually equipped so that they can use their gear for pelagic fishing – even often for herring.

While that trend to meet individual circumstances can be momentarily beneficial it would seem that to cope more efficiently with depleting stocks (some believed already at danger level) specific gears should be more generally adopted. By careful and selective concentration on shoals of mature herring some of the indiscriminate catching of young immature herring could be avoided and the possible damage done by mass hauls reduced.

Appreciation of the foregoing factors has led to the production of this book aimed at the more efficient design and handling of gears fitted for the capture of pelagic and semi-pelagic fish in European waters and elsewhere. This relates specifically to herring stocks in the European area, but also to many other stocks around the world.

Pelagic and semi-pelagic trawls are already in use or are being planned for more extensive use in a number of valuable fisheries. Many of these successful developments have contributed to the commercially sound specifications already in use. More is given about this in Chapter 1 and Chapter 3 also of the German company H. Engel of Kiel in devising early pelagic gears and assembling them.

Because of the increasing impact of quota systems, fishing rights and fishing limits the highest efficiency in catching techniques is called for in what is now a transitional phase. Research and development will continue beyond doubt, but the trend most probably will be towards 'unit-type' of fishing. Therefore the information in these pages will provide much useful reference data and the sphere covered is meant to assist in the appreciation of pelagic gear design. It may be valuable to those responsible for new fisheries. Ideas stimulated by the various chapters can easily be supplemented with precise information from manufacturers.

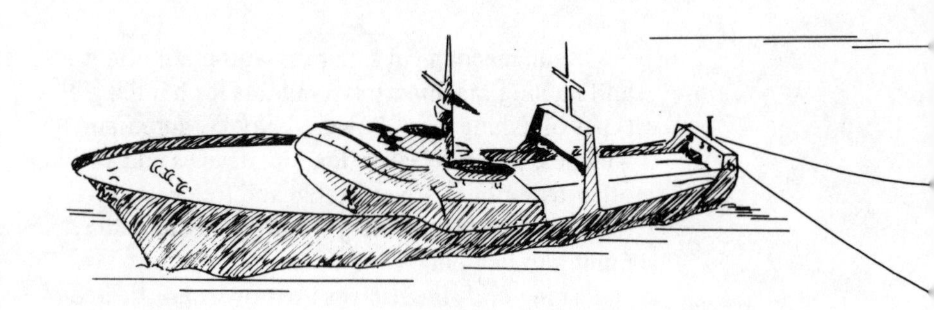

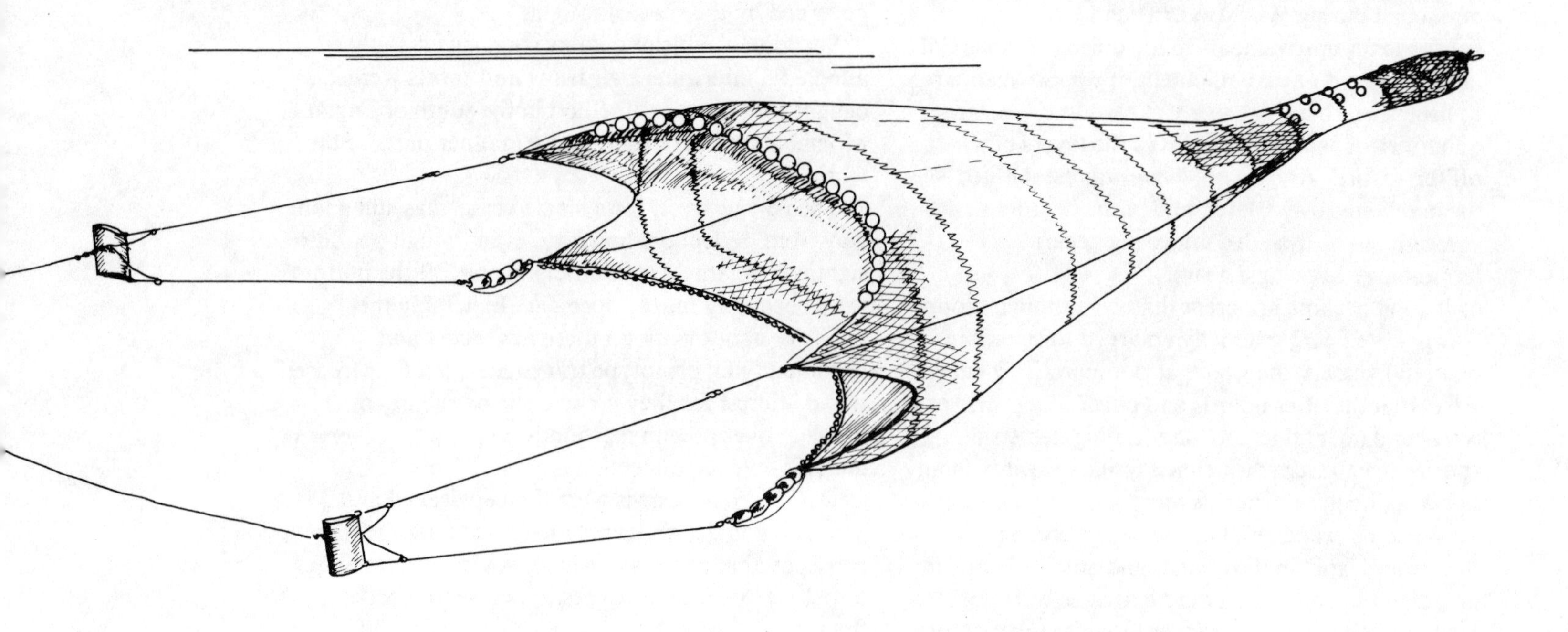

Fig 1 A pelagic trawl in operation from a stern trawler

1. The Development of Pelagic Trawls

It is highly probable that some time after the development of steam powered trawlers, fish were caught at certain periods with towed nets between the surface and the seabed, for there was a great surge of experimental net design work during the mid-1920s in Europe and probably other parts of the world.

In Britain some of the research was directed – with little success – towards cutting fuel costs (saving coal) so that vessels could go further or stay on the fishing ground longer, catch more fish and, as is always the goal, make more money.

With the aid of electronic sounding devices developed during World War II and the subsequent improvements made during commercial adaptations of naval equipment, it was soon apparent to fishermen that there were, at certain times, large quantities of fish to be caught in the layers of water off the seabed. And an objective was established – the mid-water trawl, later, and perhaps more correctly, to be included under the general terminology of pelagic gears.

It soon became apparent that even though ground trawls were being gradually enlarged to suit more powerful vessels, the gear was comparatively small. Also, that the otter boards and cover of a trawl net were fundamental in catching feeding demersal species which were disinclined to move swiftly in an upward direction, whereas most species located between the seabed and the surface were, on occasions, rapid in movement and able to escape in any direction, assuming their activities were not governed by a particular layer of water temperature.

The principal objectives therefore seemed to be twofold, either a very big net to cover a large area, or a trawl net designed to maintain a set depth at much higher speeds. In addition, it was felt that a great advantage would be gained if the depth of a gear could be monitored during operation.

Large nets were soon produced to be towed between two vessels and these achieved a certain degree of success, but equally it proved difficult to make a comparatively large trawl net which would open properly in mid-water during operations from one vessel. This aspect also pointed to the tremendous drag of large nets, and led to gradual and continual mesh enlargements. This improved the frontal areas of pelagic trawls and consequently increased the graduations in the panels. The mesh requirement of the tailpiece (or codend) was governed by the species sought.

Scandinavian fishing boats soon successfully adopted a mid-water pair trawl and sprats were caught with a similar method in the south of England. Modifications were soon in use in other parts of the world.

The pelagic trawl for a single vessel was still a long way from perfection, however, even though periodic sightings of large Arctic stocks rising 'off the bottom' were becoming more noticeable. Inevitably the publicity about work on these new gears and accompanying prototype transducers led to increased world interest for they created the possibility of trawling over potential grounds which were otherwise unaccessible to conventional gears.

Dual-purpose trawls were then envisaged and research and design teams took up yet another prospect, that of a trawl which would fish just above a rocky ground or, if necessary, close to a better floor.

A Commander Rowton who had designed and patented many types of trawls, and was later engaged in paravane design with Boom Defence during World War II, produced his 'World Master Trawl' in the early stages of this development, but to his

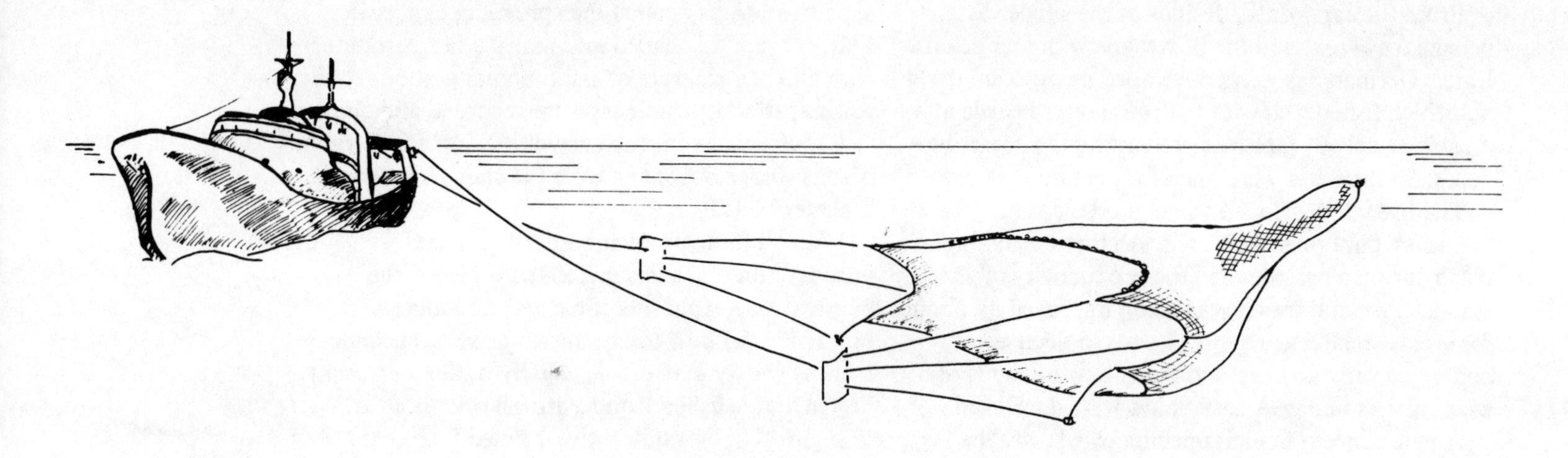

Fig 2 Dimensional sketch of a stern trawler towing a pelagic gear in mid-water

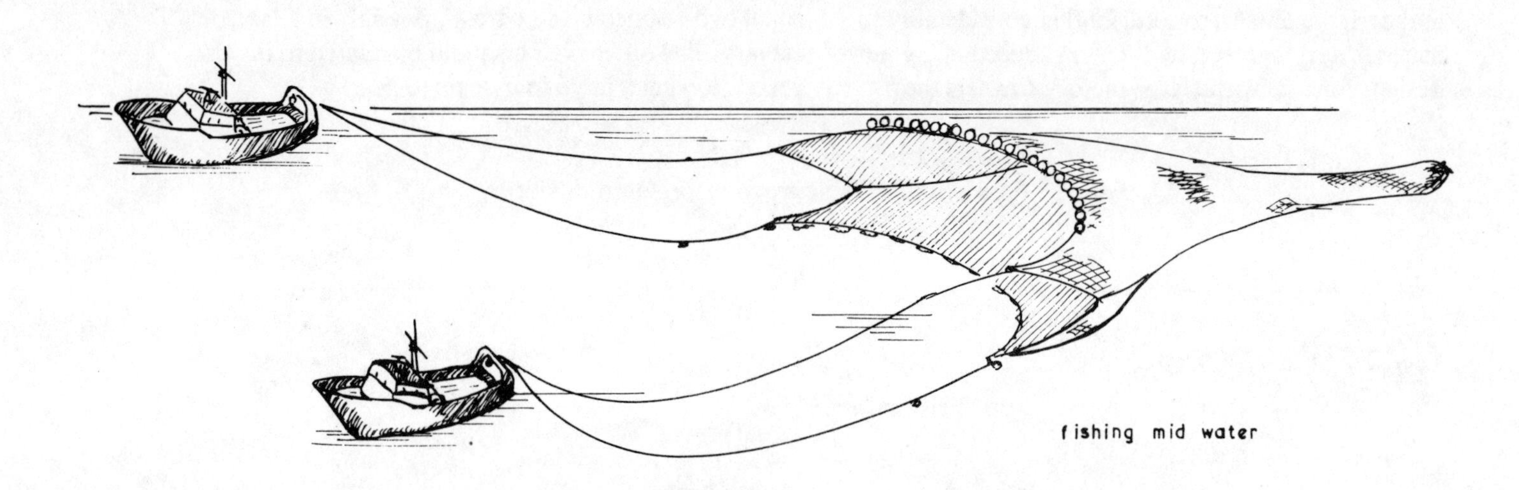

Fig 3 Small boats pair trawl fishing mid-water

disappointment, it received little support.

An early design was the gigantic American gear called the Cobb trawl. It stimulated many ideas for experiment with its curved otter boards and ingenious assembly. A forerunner, the Icelandic floating trawl, was being continually modified.

From the early days, designs of the single vessel pelagic trawl featured the box shape with four seams. Later, German designers developed a two-seam trawl net which fished well with hydrofoil otter boards and a cable attached transducer recording the operating depth. So the gates were open for definite progress.

The French developed some good plans of two-seam dual-purpose trawls. All this was taking place during a period when the productivity of standard ground trawls was being improved by aiming for greater headline height. The two objectives combined satisfactorily and many new methods of shaping nets and gear assemblies were developed.

A new concept in high opening trawls was the wing trawl which was introduced into Britain from Scandinavia in 1958. The whole new concept of this net lent itself to the idea of 'off the bottom' fishing, and it was soon adapted to fish mid-water with conventional otter boards, and the assembly spread to other countries.

During these post-war years the Japanese research and design teams were contributing considerably to fisheries development, and they produced many new designs for use by the large factory trawlers they were then building.

Numerous other ideas were mooted such as dimensional gears with four or even six otter boards, or a combination of otter boards and paravanes. And research in test tanks was directed towards otter board design, with concentration on the shape and aspect ratios, as against the spreading capacities. Flat, curved, 'L' shaped and many other variations with different degrees of slant, and alterations to the towing point attachments were recorded and charted.

Requirements for the correct loading and flotations of nets were resolved as were the snags of releasing the otter boards.

With all these problems solved and the introduction of transistorised transducers, the cumbersome expensive cable was no longer a hindrance. So now today the single vessel pelagic trawl is widely in use. It is usually of the four-seam design and is fished with hydrofoil otter boards.

Figure 1 shows how a two-bridled four-seam pelagic trawl gear might appear in operation from a large stern trawler, whilst Figure 2 gives an illustration of a small three-bridled two-seam gear being fished from a medium class stern trawler.

The pelagic pair trawl operation is depicted by Figures 3 and 4. The first drawing represents two comparatively low powered open boats towing a small two-seamed pelagic trawl, possibly in a lake or estuary. The other sketch should be related to larger vessels towing a large four-seamed gear.

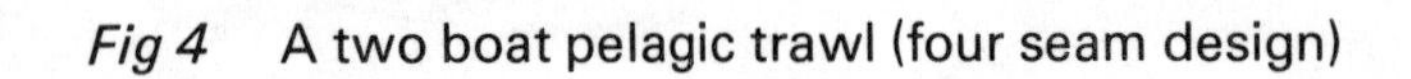

Fig 4 A two boat pelagic trawl (four seam design)

2. Examining Existing Net Patterns and New Concepts

The sketches Figures 1 to 4 in Chapter 1 illustrate how pelagic trawls might look in operation. They show in each case voluminous net designs which have been made for towing, with correct loading and flotation by (1) either a single vessel gear held open by using hydrofoil otter boards (or as explained later, in the case of some small gears, specially adapted rigs may be fished with conventional or 'V' form otter boards), or (2) the pelagic pair trawl method where the vessels maintain the fishing 'spread' of the assembly during the towing operation.

For those readers who may not be very familiar with pelagic trawling principles, Figure 12 on page 18 represents an inshore vessel towing a pelagic net in mid-water. The major parts of the assembly are named for the guidance of the reader.

Variances in Pelagic Net Patterns

There have been numerous innovations of net patterns for improving pelagic trawls, and in this chapter it is proposed to examine briefly the differences, also to look at possible advantages and disadvantages of the existing designs.

As has already been explained, initially there were two basic patterns, the two-seam net and the four-seam (or four-panel design). And it now appears, although it was not always so, that the four-seam version is preferred for the larger vessels such as factory trawlers, whilst the two-seam type may be preferred for working from the inshore or medium class of vessel.

Similarly with pelagic pair trawls, they may be of the two- or four-seam variety, and again it is probable that four-seam nets are operated from larger vessels, whilst the two-seam type are preferred for operation from smaller open boats.

On examination these preferences may be understandable to some extent, if one considers the possible 'drag' ratios of the different net designs. It is not proposed to give any scientific theory in this book, however, as this sort of calculation can be left for the Marine Laboratories to expound, although it is intended to illustrate by examples how these matters may affect preferences for precise patterns. Also, to some degree, help the reader to determine which pattern may be the best for him to use.

It is important to understand that from the early stages pelagic net design patterns were very large and consequently subject to large drag factors. To minimise this drawback the meshes in the frontal area, such as the wings and square, were comparatively large, and the panels down the body of the trawl decreased proportionately to meet the tailpiece, which was of a mesh size suitable for catching a specific species.

The fear was that if the mesh size in the frontal section of a net was too big certain species might be allowed to escape. This is not so, and in fact certain nets now feature extremely large meshes in the wings and leading panels of the trawl. This has been part of the evolvement of heavier net machinery. When the 35 mm machine was designed, the intention was to produce heavy codend material for stern trawlers. Now these machines can be employed for making the large mesh sections required for pelagic trawls, using materials which are approaching fine rope sizes rather than twine.

Figures 5 and 6 show the sections of two- and four-seam pelagic designs. These are quite explicit in themselves, and if studied will help to illustrate the following examples.

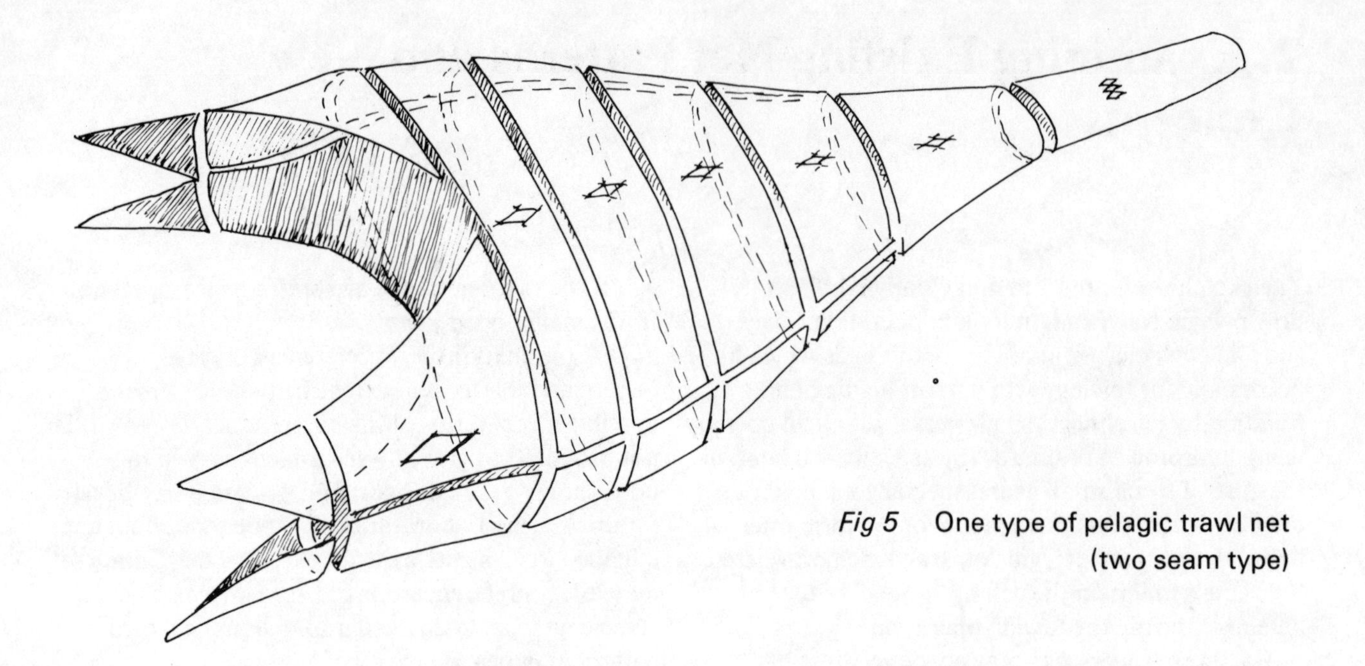

Fig 5 One type of pelagic trawl net (two seam type)

Fig 6 A perspective view of the sections of one type pelagic trawl net (four seam design)

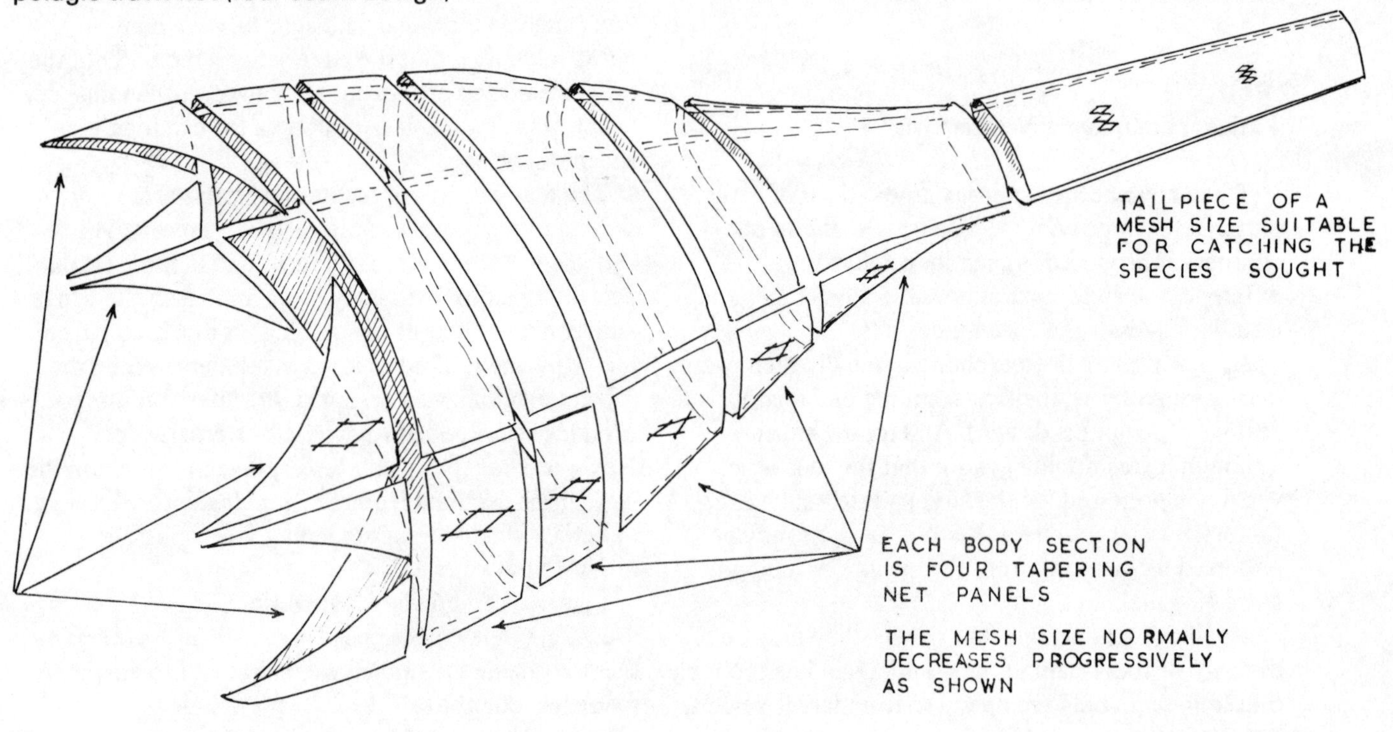

Figure 7 illustrates pelagic trawl net patterns (examples A and B).

These examples show the forerunners of pelagic trawl net patterns. Example A shows how two identical sections are seamed down from X leaving the 'V' shaped wings. This is possible because of the type of cut used for making the four wings. And the perspective inset further illustrates this.

It was designs similar to this, but with extended wings, that were used in the early days for making trawls for fishing mid-water from all sizes of vessels.

Nevertheless it can be seen that the gape of the trawl is somewhat restricted, and this would also be the case even with the wings extended to leave 'V' tips.

Looking at Example B, the mouth area is considerably extended with the four-seam design (or four-panelled net) and that gain is achieved by the particular shape of the wings. The 'V' at X allows a somewhat improved vertical opening.

Figure 8 illustrates pelagic trawl net patterns (examples C and D).

The first drawing (C) shows a different kind of four-panelled net, where the top, bottom and sides are roped equally. Thus two of the panels become the top half of a trawl whilst two similar pieces make up the lower half. This design is obviously more suited to the smaller type of vessel and for pair trawling; again the improved 'V' at X gives adequate vertical opening.

The second example (D) is a variation with a gusset inset, more than a fully four-panelled type. The wings being straight on the outer edges combine with the particular shape to present a net with good spread and reasonable vertical dimension because of the larger opening at X.

Figure 9 illustrates pelagic trawl net patterns (examples E and F).

The examples here describe shapes on which the larger pelagic trawl net specifications may be designed. They show quite voluminous assemblies,

Fig 7 Pelagic trawl net patterns

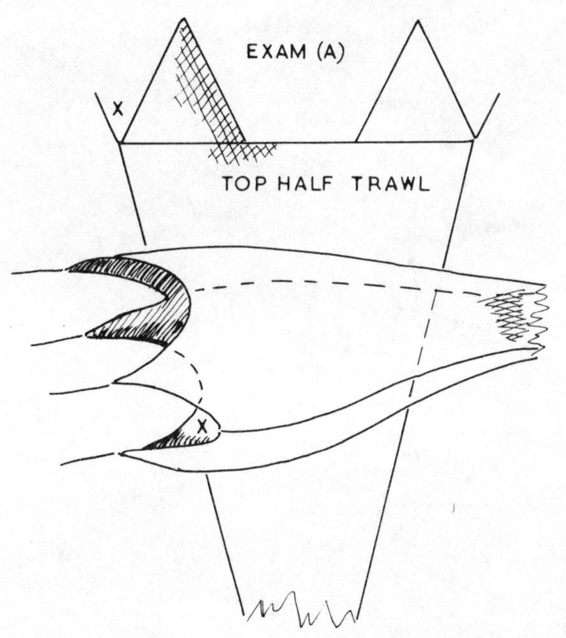

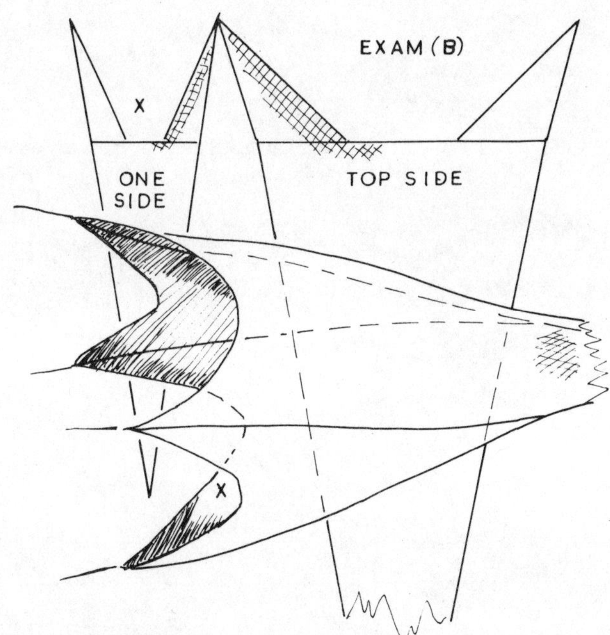

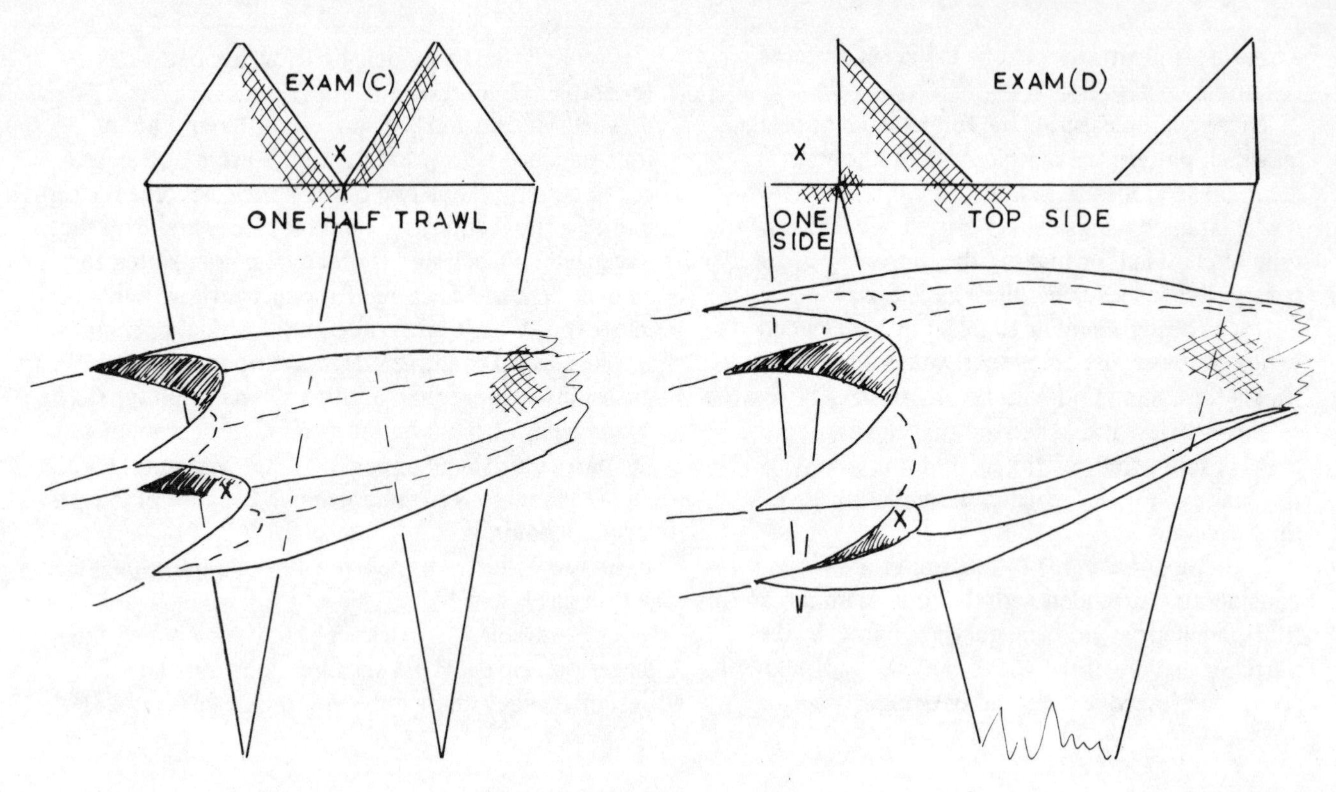

Fig 8 Pelagic trawl net patterns

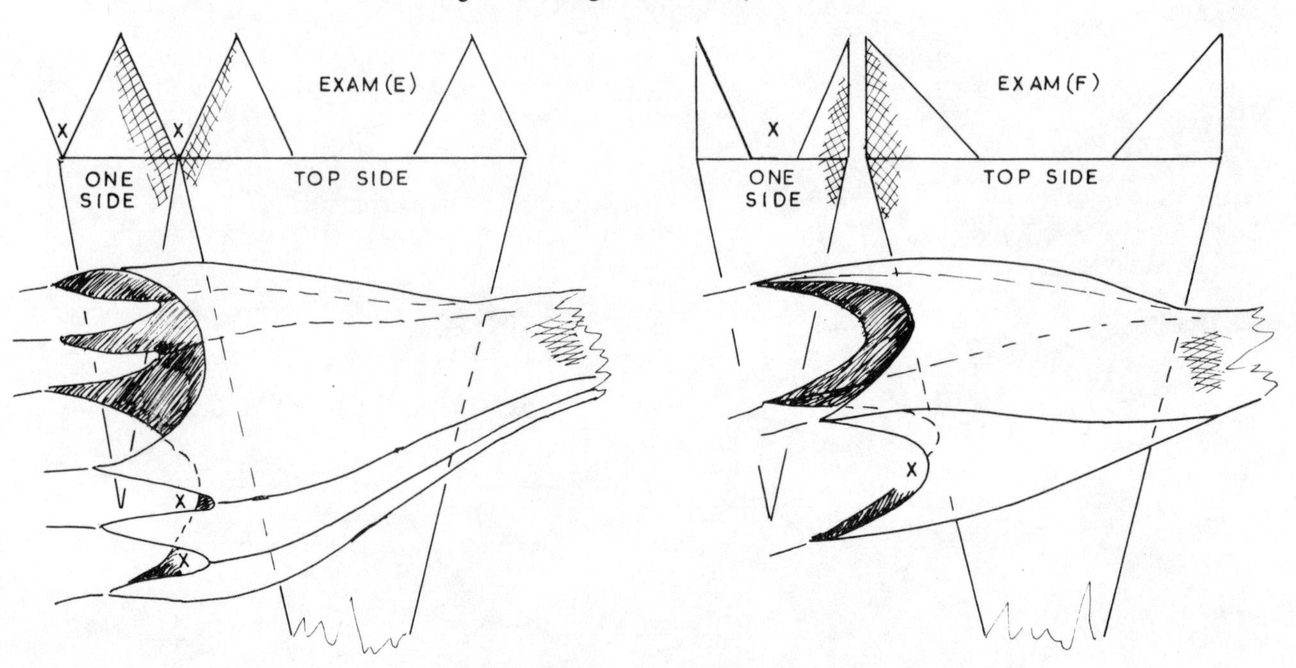

Fig 9 Pelagic trawl net patterns

and in each case they are four-seamed examples. The length of the wings may be far more extensive than is indicated on the sketch.

The former outline (example E) shows a major variation on other types, with two double 'V's being formed at the position marked with an X. These are to facilitate a three bridle gear.

On the other hand, the second sketch (example F) is similar to example B but it is actually a design preferred for nets used by large fishing craft.

Possibilities for Improving Pelagic Trawl Net Patterns

Some of the existing shapes of pelagic nets have been examined in the foregoing, and it is now proposed to look at some other possibilities for improved shaping.

Before proceeding, however, it is most important to mention that one of the major factors governing designs of net specifications relates to the planning of the manufacture of the sections, and the assembly factors. A second consideration is that of repair after damage. (Although in fact, pelagic gears are susceptible, except during operation in bad weather, to only slight wear and tear.)

Figures 10 and 11 should be considered in conjunction with one another, for they are meant to illustrate present restrictions to trawl mouth opening, and show possible ways of increasing the horizontal and vertical 'spread'. For the purpose of illustrating this, the fundamental feature, each design has been shaded, *ie* the main perimeter of the body netting.

From these sketches it will be seen that the two-seam net is restricted vertically, whereas the four-seam pattern has a greater vertical opening because of the side panels. The six-seam net (suggested

Fig 10 Pelagic trawls (two and four seam designs)

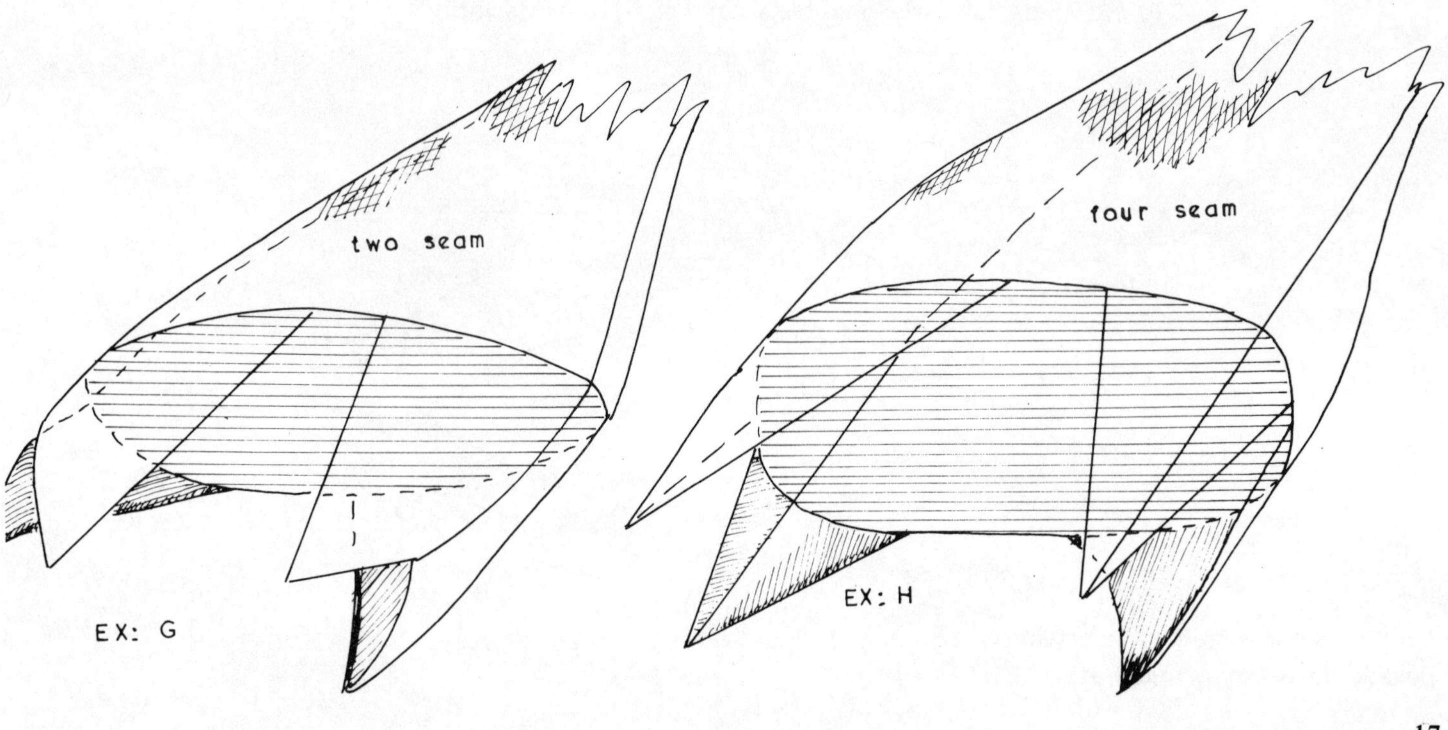

two seam
EX: G
four seam
EX: H

pattern, example I) achieves greater dimensions, and this is more emphasised in the further idea of an eight-seam trawl net which is described by example J.

Note: It is normal to run supporting lines down each seam, as well as crosswise or laterally and down the quarters.

The size of these ropes and their number depends on the dimensions and bulk of the particular specification, and these details are normally available in full from manufacturers.

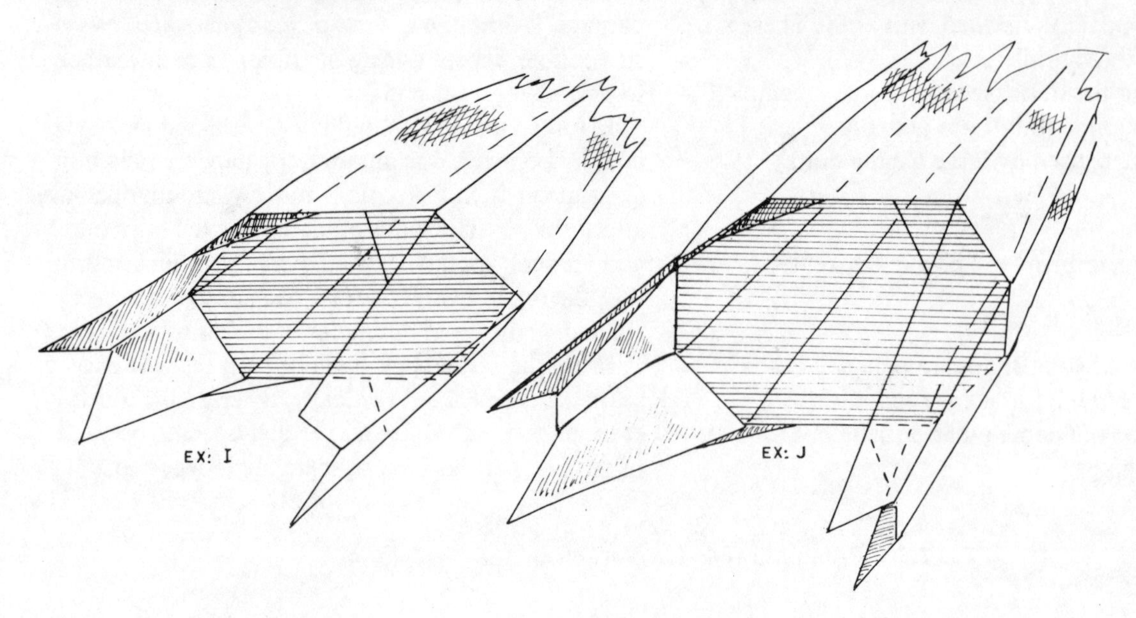

Fig 11 Pelagic trawls (possible six and eight seam design)

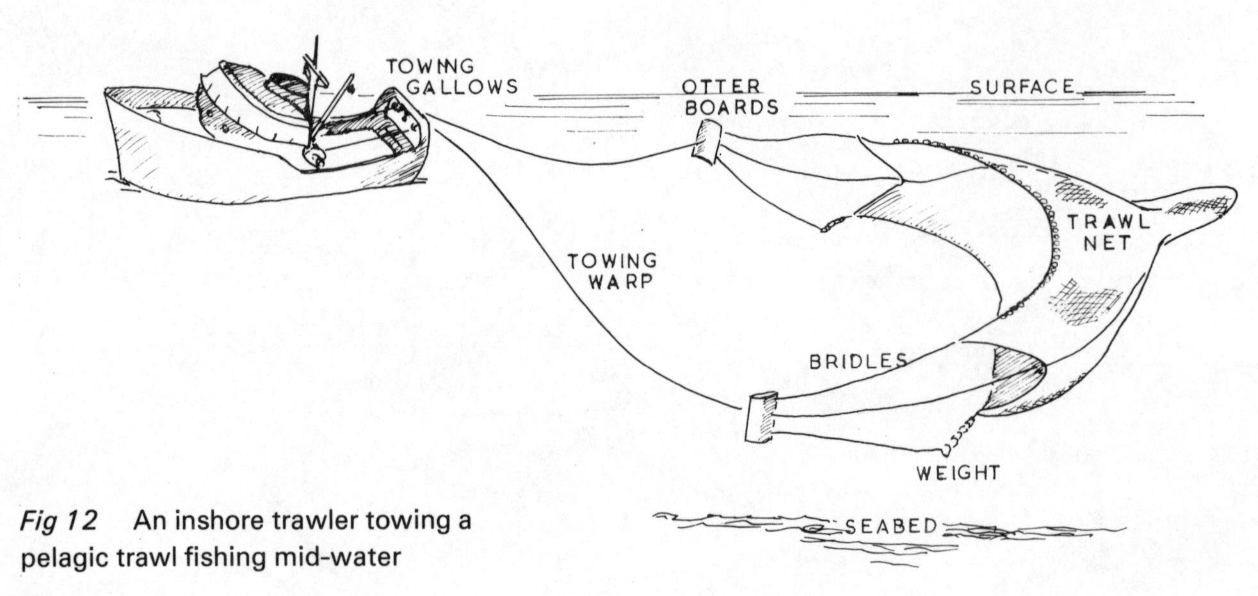

Fig 12 An inshore trawler towing a pelagic trawl fishing mid-water

3. Specifications of Single Vessel Pelagic Trawl Nets and Different Towing Assemblies

The text so far has covered the evolvement of different main patterns for achieving the best openings and dimensions of large pelagic nets. The obstacle to be overcome was 'drag' or the resistance of water passing through the netting. Obviously the smaller the mesh the greater the resistance. For example, a sheet of netting with minute apertures will be more restrictive and cause heavier load factors than a panel of netting of similar dimensions but with large meshes and of comparable twine size. Equally, the heavier the twine the greater the drag forces. The objective, therefore, is to achieve a balance.

With this established the emphasis was gradually to increase the mesh size in the frontal parts of a gear using correspondingly heavier twine. The follow-up panels were consequently subject to mesh size increases, with the 'joining-up' ratios normally maintained, that is to say, where the tapered width of one panel meets the forward width of the next, the total measured stretched lengths are normally comparable. (Example: 200 meshes × 6″ mesh = 300 meshes × 4″ mesh.)

At this stage it should be mentioned that this reasoning should be applied with reservation. The principle is often used when designing many different types of nets, and it is one that should assure maximum 'flow' to the netting of towed gears. It must be remembered, however, that the 'drag' build-up does create high forces within and around the perimeter of the body netting, and that these factors do naturally cause increasing distortion in the narrowing tunnel of the circular-shaped body netting as the meshes contract under the mounting forces.

It is often advisable, therefore, indeed usual with large nets, to use a differential ratio, where the mid-to-tail panels are designed marginally (10% to 15% or 20%) less in width than the one before. And when the meshes are joined proportionately a more even flow can be obtained. The percentages can be graduated down the body netting for the best results, and this is where tank test facilities can be most useful.

The percentages mentioned above are not always incorporated in more recent designs. And in fact a 'blow out' section may encompass the trawl net in the body of the second panel. It is of extremely large mesh netting so that drag is minimised and flow of the netting is maximised. (Figure 13.)

Nylon twines have been found to be the most suitable for pelagic trawls, and fortunately machines for manufacturing the 'cords' needed for the newer designs have quite recently been developed. Polypropylene or polythene materials have some success in use for semi-pelagic trawls, but not for nets principally designed for mid-water to surface fishing. Terylene type ropes are preferred, as many net patterns are also based on percentage allowances of netting to ropes for increasing the fishing gape or dimension. Rope stretch is undesirable.

The side trawler represented by Figure 14 shows a vessel part of the way through hauling a pelagic gear, with the otter boards at the gallows and the net to be drawn in. This illustration is added, as so far only stern trawlers of different categories have been shown towing pelagic gears, and it is not unusual for side trawling vessels to operate mid-water techniques, particularly of semi-pelagic assemblies. The voluminous netting, however, can be a problem when worked from a side trawler, for it has to be specially rigged; whereas the natural deck space, hauling lengths and flaking amenities available on stern trawlers makes operation from them far easier.

Fig 13 A pelagic trawl with large mesh 'blow-out' panel for easy flow in mid-water

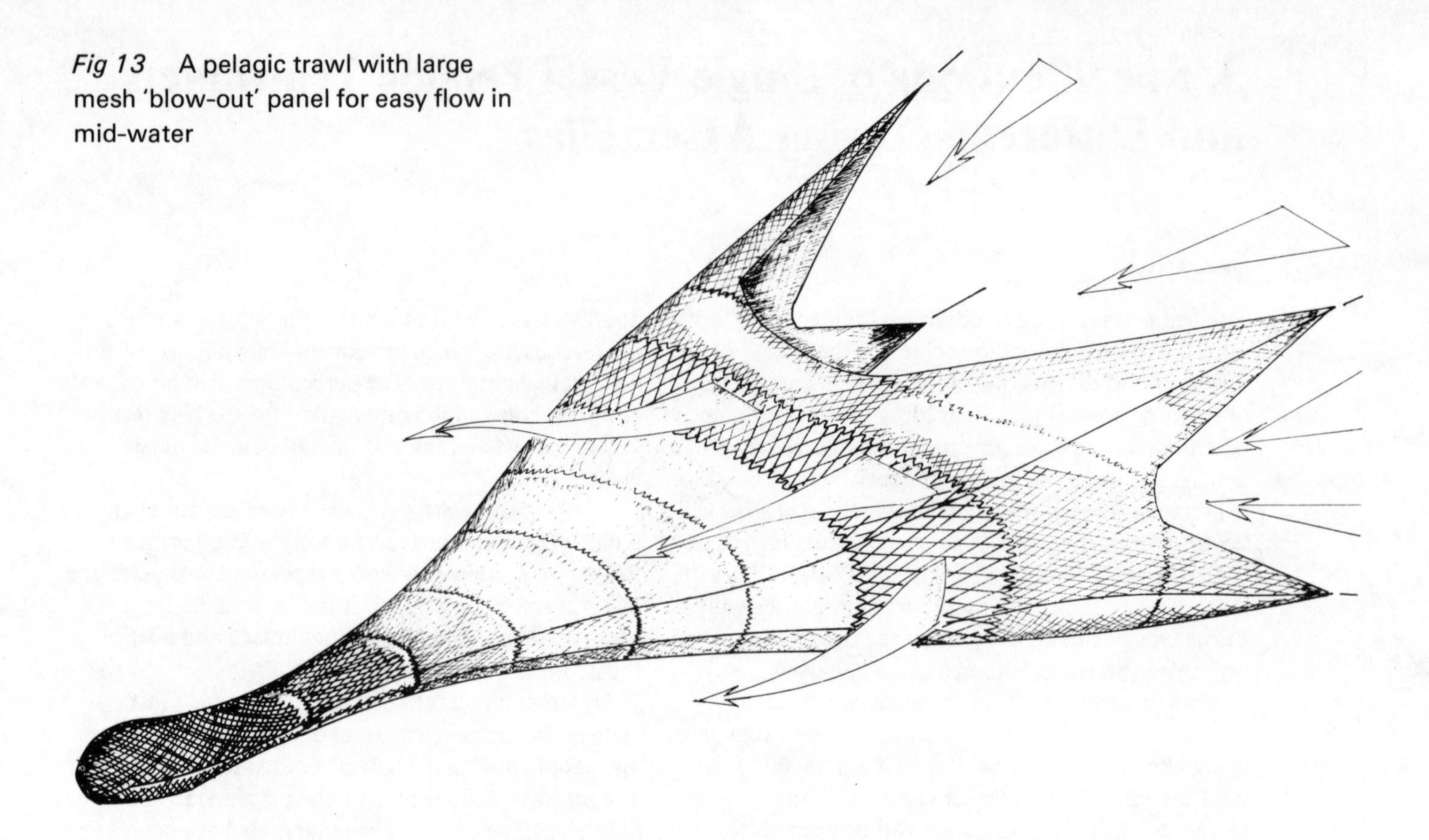

An Early 1960s Commercial Two-Seam Pelagic Trawl for Single Vessel 800/1000 h.p.

The first specification given is one of the early pelagic gears to be operated with some commercial success. Developed and made by H. Engel of Kiel, it won a good deal of comment as the first real advance in fishing the mid-water levels of the oceans. There had been earlier major pelagic gear successes before this, such as the Danish Atom trawl in 1948, but these, although forerunners of improved designs, appeared complicated, and were more appropriate for use in fishing coastal waters rather than oceanic waters.

The design featured by Figure 15 was for a deep sea vessel, but the manufacturers issued a range of gears to accommodate different types of craft, including inshore vessels. And these designs were all of the two-seam type.

A Modern Four-Seam Pelagic Gear 1000/2000 hp Vessel

The specification depicted by Figure 16 is that of a modern pelagic trawl in use from deep sea vessels. The trend, as has been explained previously, has been in favour of four-seamed net patterns. And a feature that recurs with all fishing gear specifications is that there are always minor variations and small preferences in design. Manufacturers are often dogged by having to record these preferences of particular customers, and it is true that some of them hardly do seem worth the trouble. The facts are that once a particular assembly has been proved successful in operation the basics of design are established, and these are the principles sought to be related in this book.

Some differences are worth noting. For instance, in the early days of mid-water assemblies, or floating trawls as they were then known, it was not certain whether the main cover (square) should be featured uppermost or along the lower edge of a net.

Eventually the top and lower sections were made equal, and the side panels were made to match in length but with varying depth ratios.

Four-seam nets are made for all ranges of vessel to suit particular horsepower capacities, and they may have a lower cover or top cover, but these are usually only marginal overlaps.

One Vessel Pelagic Trawls as Subsidiary Equipment

It is often found that pelagic trawls are regarded as subsidiary equipment rather than everyday gears.

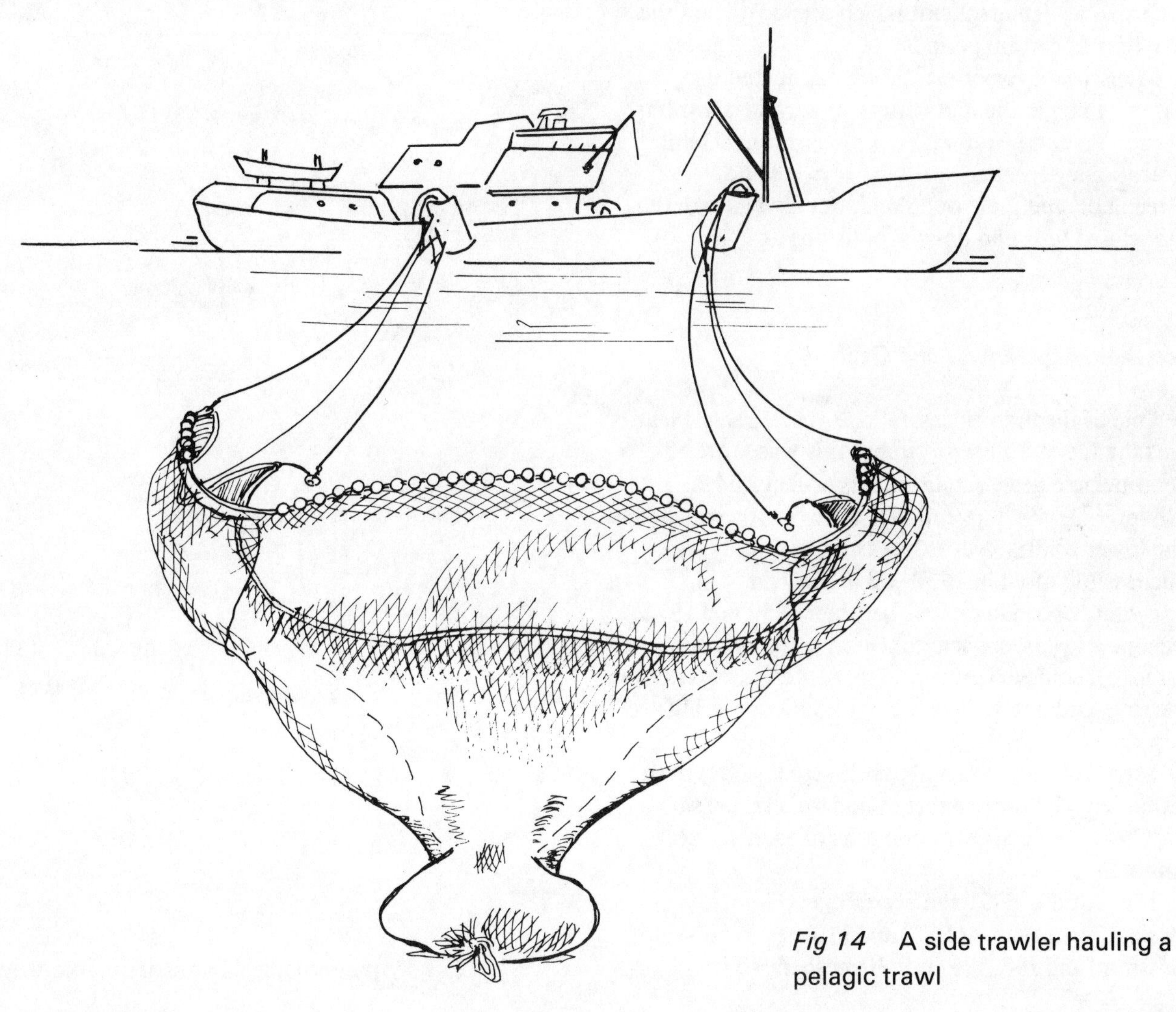

Fig 14 A side trawler hauling a pelagic trawl

For example ground trawling or demersal seining can be conducted all the year round, weather permitting, whilst it is often the seasonal spurt of pelagic species that call for mid-water towing operations. The economics of fishing middle water layers with trawls therefore can be precarious, unless the activity is specific at certain times of the year when fish behaviour can be anticipated. Alternatively, pelagic trawling can be a valuable substitute operation.

Large stern trawlers built to fish ground gears now have modified ancillary equipment for using a pelagic assembly as the opportunity arrives. Such an occasion may be to capture species normally regarded as demersal, but which are located off the sea floor at certain periods.

Much of the experimentation conducted for catching fish in the mid-waters-to-surface areas has been in northern waters. Subsequently successful operations were made off Norway, Iceland, Greenland and Newfoundland, but always with the associated 'ups and downs' of fishing.

Icelandic Experiments and Gears

One of the first successful one-vessel pelagic trawls was the Iceland *Breidfjord*, but this was more of a semi-pelagic gear, similar to a type featured in Chapter 5, with conventional otter boards attached to the lower bridles. Nevertheless the assembly was successfully used in 1952–58 in the fishery of spawning cod. Thereafter, between 1959 and 1972 pelagic trawls were scarcely used commercially, although some experimental work was carried out on herring, cod and redfish, but without convincing results.

More recently cod and capelin stocks have been exploited with more effect in mid-waters, possibly catching above 4000 metric tons of each species annually.

Most of the nets have been Engel (Germany) designs in the ranges of 572 and 710 mesh trawls (of 560 mm) and 398, 506 and 630 mesh trawls (of

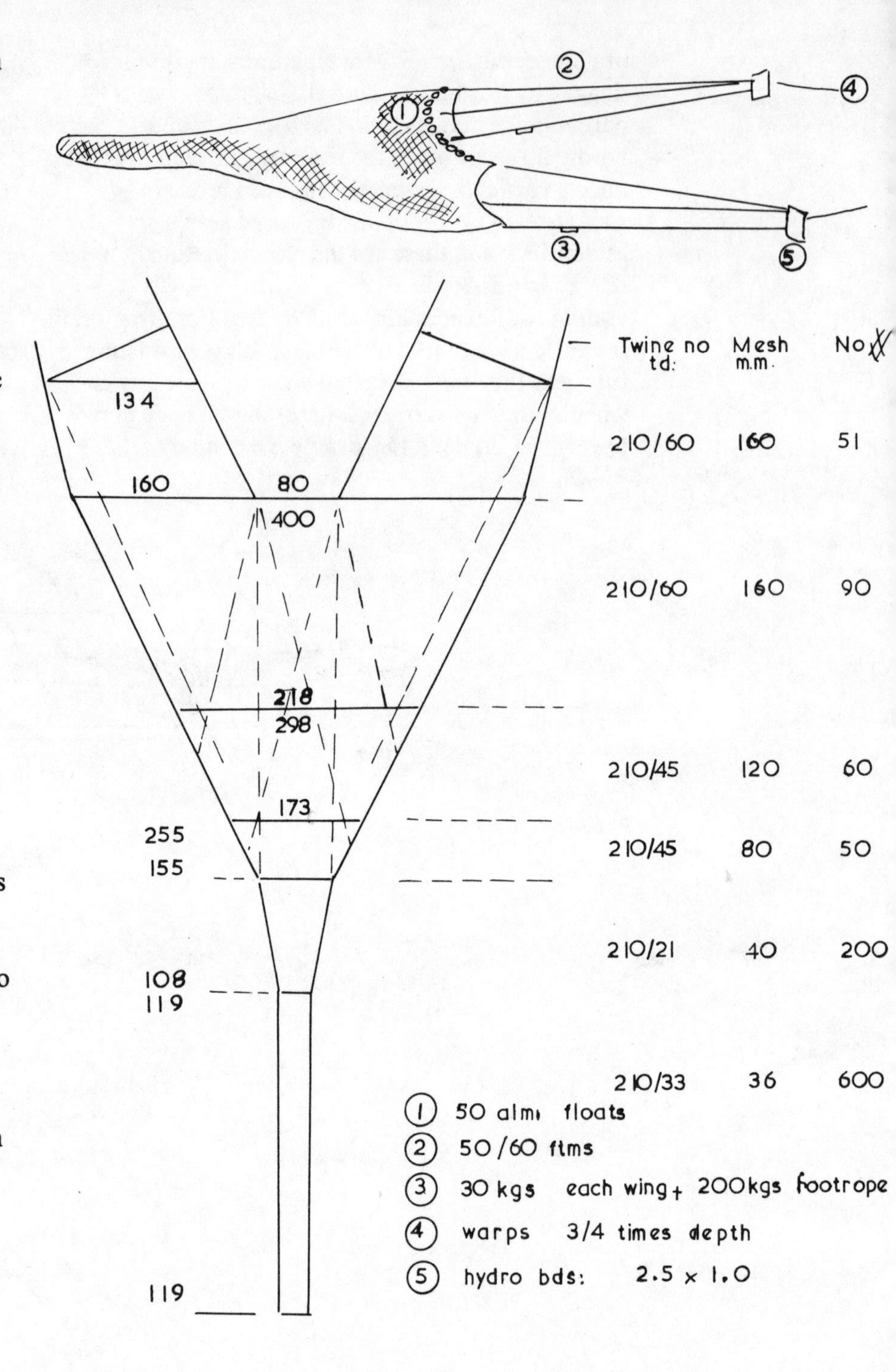

Fig 15 Two seam single vessel pelagic trawl (800/100hp)

800 mm mesh opening), using Suberkrub otter boards of 6 m^2 and possibly 8 m^2. Or for catching capelin, Norwegian trawls of the 12 × 12 fathoms, 14 × 14 fathoms and 16 × 16 fathoms are used.

Inshore One-Boat Pelagic Trawling

The drawing, Figure 17, shows an inshore vessel towing a pelagic trawl. It also clearly identifies the main net sections, and indicates a two-bridle assembly with a four-seam net. Two seam trawls may also be operated as efficiently with three bridles, the middle bridle being so balanced that it carries the main weight from the side lines of the net and permits the netting to open to the best vertical advantage, with a predetermined 'spread'. A two-seam net may, however, be worked with only two bridles, for a great

Fig 16 A 1600m pelagic herring trawl

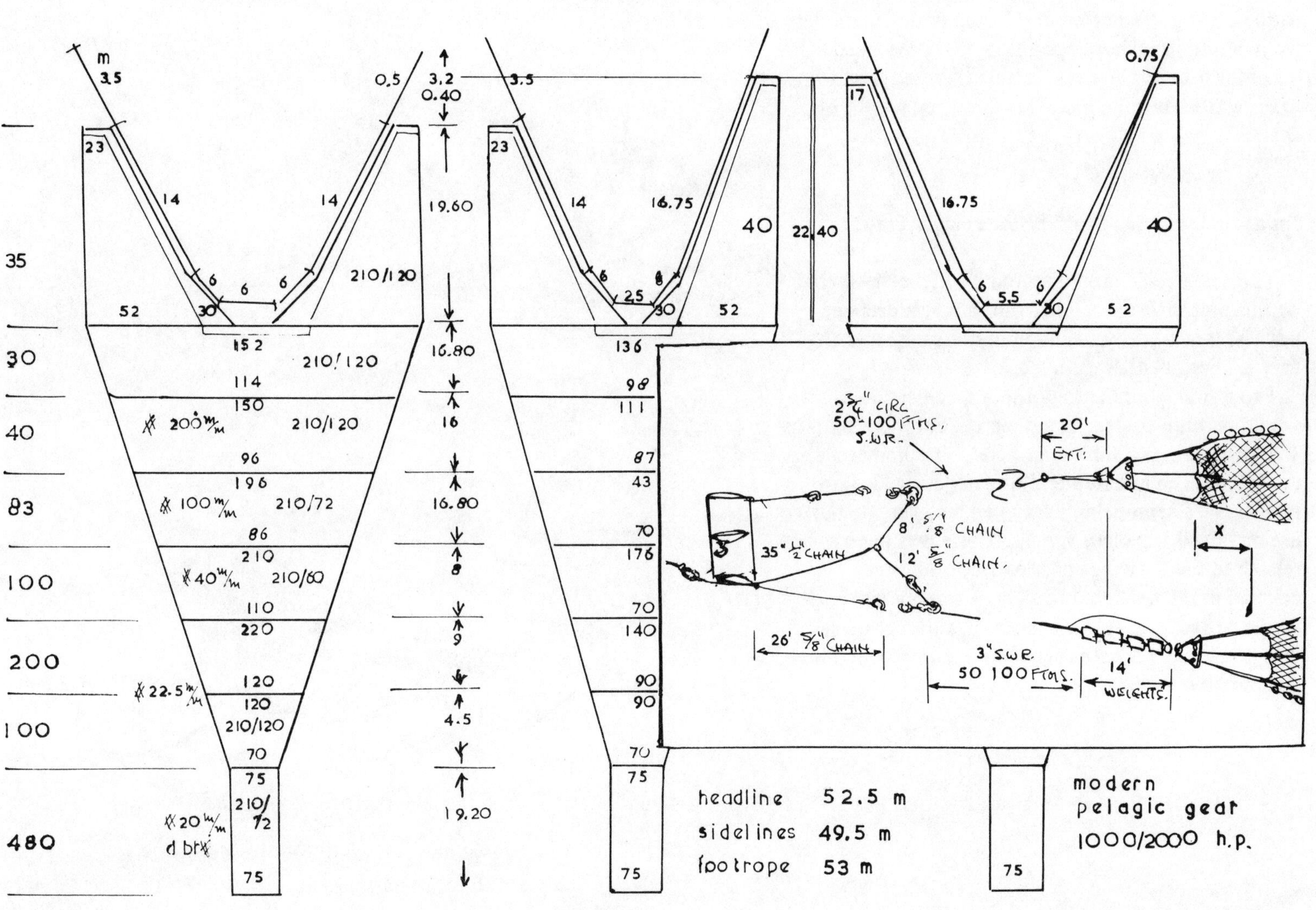

deal depends on the particular releasing assembly of the gear in use.

300/400 h.p. Canadian Four-Seam Pelagic Trawl

A further specification is described in Figure 18. It gives details of the net plan and assembly of a four-seam net suitable for near-water vessels. Worked with hydrofoil otter boards the panels are so arranged as to give equal openings at the top, bottom and sides, or a frontal rectangular aperture under towing conditions. This sort of design is not to be regarded as the usual preference, but the idea of designing nets for medium or inshore operation has some merit. The benefit is that it is less susceptible than some patterns to distortion through incorrect flotation, weighting, *etc.*

Japanese Deep-Sea Multi-Seam Pelagic Trawl

The net specification featured by Figure 19 is that of a net pattern in operation from Japanese deep sea vessels. It is an extremely voluminous net requiring four bridles at either side.

From first examination of this specification it would perhaps appear complicated, but it is a pattern well proved in test tank and thereafter commercially. As will be seen in Chapter 5, the net design is fairly similar for a semi-pelagic trawl which is in popular use on board Japanese vessels. These nets have to be related to the drawings of Japanese otter boards covered in the next chapter. The assembly and release would be basically as that described, although there are variations and there are particular features for hauling on board by stern trawlers.

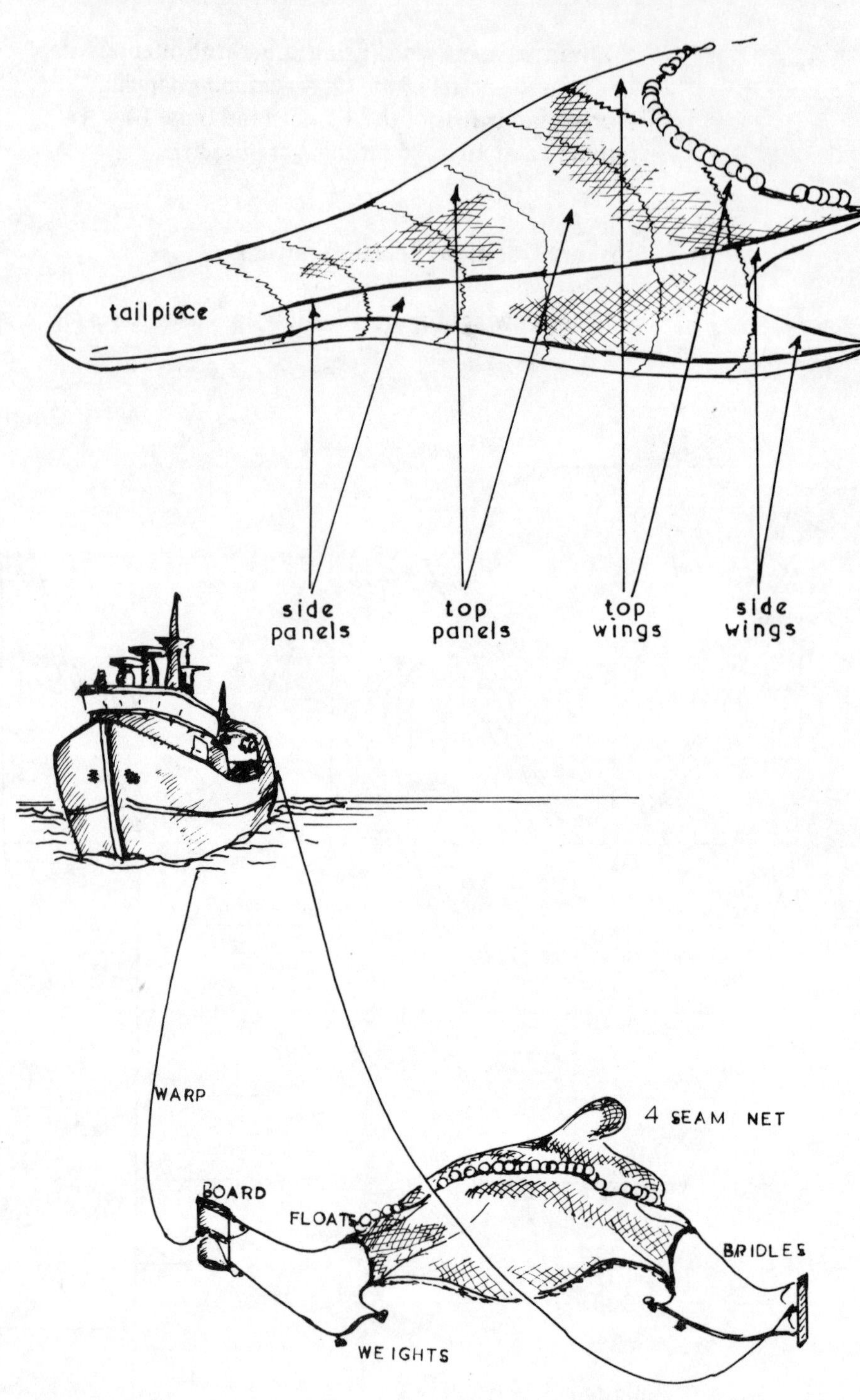

Fig 17 Trawling mid-water and identifying sections of a small four seam net

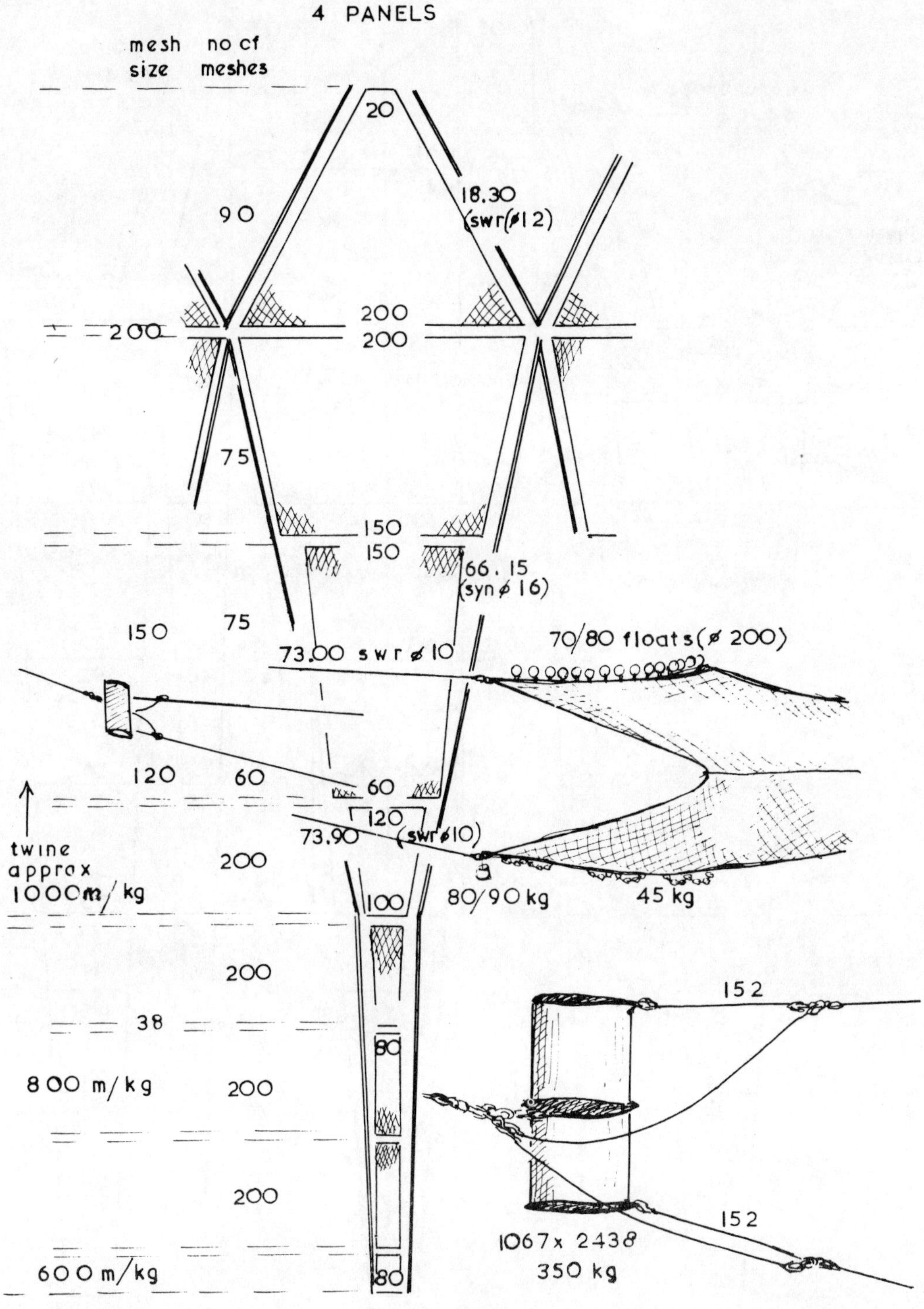

4 PANELS

mesh size

no of meshes

20

90

18.30 (swr ∅12)

200

200

200

75

150

150

66.15 (syn ∅16)

150

75

70/80 floats (∅ 200)

73.00 swr ∅10

120

60

60

120

73.90 (swr ∅10)

twine approx 1000 m/kg

200

100

80/90 kg

45 kg

200

152

38

80

800 m/kg

200

200

152

1067 x 2438

350 kg

600 m/kg

80

Fig 18 Canadian pelagic trawl (metric), 300/400hp vessel

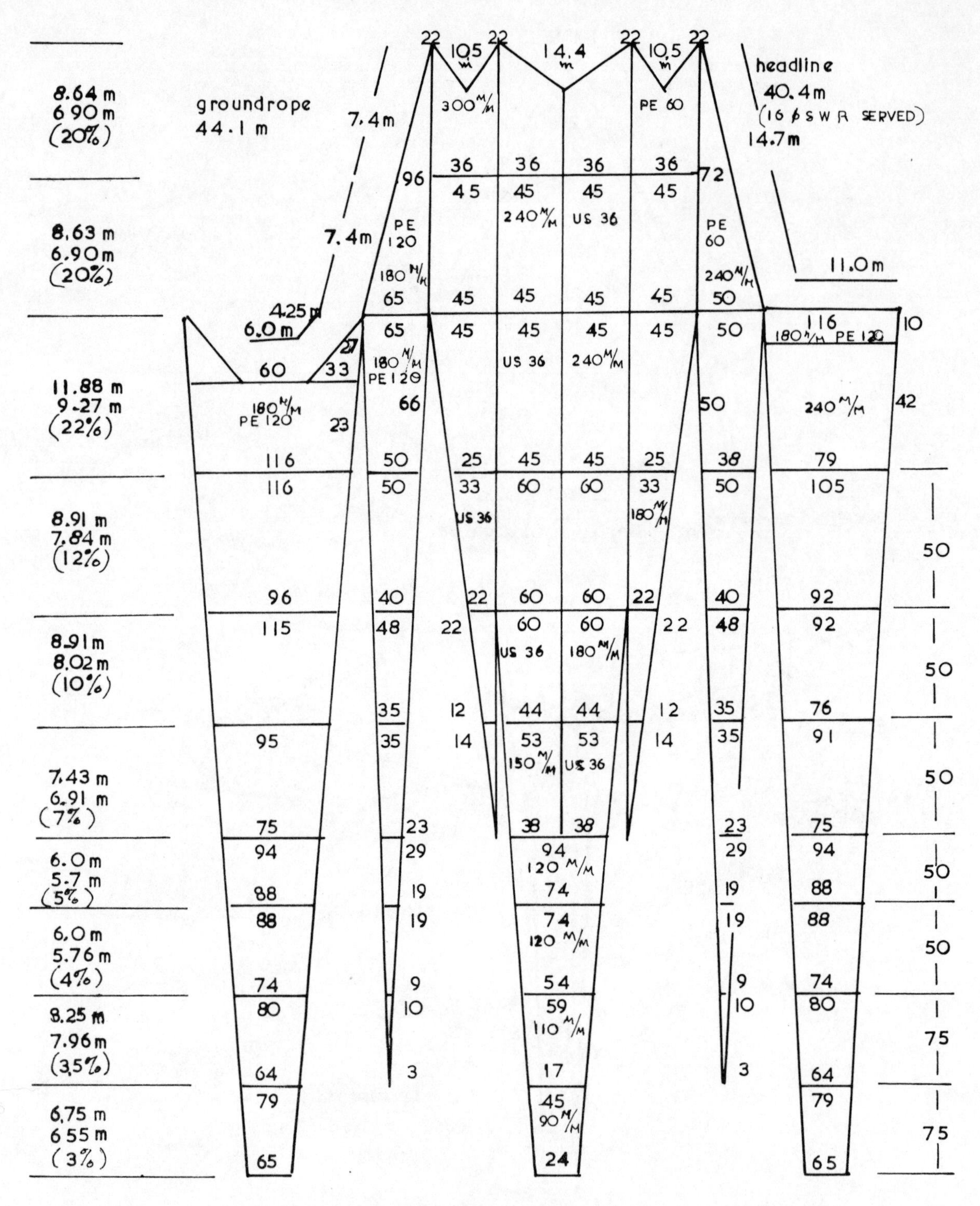

Fig 19 Japanese deep sea multi-seam pelagic trawl

4. Otter Boards and Assembly Procedures for Pelagic Gears

Pelagic Otter Boards

Otter boards did revolutionise trawling in the 1920s. Since then it has been an accepted fact that the boards or 'doors' as they are sometimes known, are fundamental to the efficiency of the trawling assembly as a whole. This is particularly so in the case of one-boat pelagic gears where the performance of the otter boards is of vital importance.

With conventional boards, the main factors are the weight and the 'angle of attack' determined by the settings. For ground trawling, provided these factors are equal for the port and starboard boards, the gear will function, even though the dimensions may be above or below the ultimate requirements. This is because ground contact is involved as a stabilising factor, and it is an important factor which cannot be considered when designing the boards for trawls to fish mid-water.

Otter boards for pelagic trawl gears do in fact need more precision to attain any sort of fishing results. The hydrofoil formation of strengthened curved steel with adjustable trawling fastenings and an aspect ratio of 2:1 (the height being the greater) was found to be the most suitable, and is now generally accepted in European fisheries.

The makers of Suberkrub define their hydrofoil boards in metres squared, and the size of boards can be related to a specific net, with assembly information to suit vessels of certain engine horse power capacities.

During the exploratory pelagic trawl gear research period a great deal of preliminary work was conducted by marine laboratories in various countries in the search for improved otter board performance. Extensive test tank experiments in Britain covered every shape possible and the staff of the Marine Laboratory in Aberdeen, Scotland, have published some excellent analyses on otter boards.

In fact, there were, and are, variations to the hydrofoil design. In each case they are of all-steel manufacture, and include modifications of many shapes and sizes. For instance, there is one pattern which is a mixture between an oval otter board and the ordinary rectangular type, with the top and lower edges angled out. But the French Polyvalent otter board is now proving to be very popular for its versatility.

Despite these variances the commonly accepted pattern for commercial use is that of the hydrofoil board as shown by Figure 20. It is of a board with three bridle connections, which have to be released from the back strops during hauling, after the otter board has reached the gallows. Release assemblies can vary, but in this instance example (a) (which has been enlarged) shows the back strop A holding the weight at the jamming link. Once the board arrives at the gallows and has been secured the G-link at Z is unfastened, and the weight is once again taken, hauling in B the independent piece, which is followed by C the bridle, thus releasing the weight on the door through A. Thereafter A is unclipped from X and the gear is allowed to be hauled in free of the otter boards. A similar action takes place at each of the other bridles, whether two, three or more.

There are always variances, for example hook assemblies, and normally on heavy gears the lower back strop would be of chain to give more balance to the otter board.

Japanese fishery scientists were early in the research field of pelagic trawl gears, and they were able to give conclusive results with their use of efficient test tanks. One of their lines of development has settled on designs quite different to those which have evolved for European waters. This features semi-pelagic and pelagic otter boards with an aspect ratio of 3 : 2 (see Figures 22 and 23).

Semi-Pelagic Otter Boards and Conversions

Whilst new fishing gears, or techniques, may often prove to be successful, it is not unusual for a considerable period to elapse before they are generally accepted. A case in point in recent years is that of the all-steel 'V' form otter boards for ground trawling in place of the ordinary wood and iron boards. A few years ago at each port one saw maybe one or two vessels out of a fleet using 'V' form doors, then they would be back on the quay. But today, perhaps because correct sizes for boat capacities have evolved, what was the conventional otter board is becoming a thing of the past.

Nevertheless, with semi-pelagic gears such as those examined in the next chapter, it is often preferable to have a heavy board with good ground contact whilst the net floats several feet above the seabed. In some instances all-steel oval otter boards may be even better for achieving the desired results, or alternatively a size higher than those of the 'V' form boards normally operated may be necessary.

The 'V' form otter board is much more versatile than the ordinary rectangular type, and it may be that its full potential for fishing off the bottom is not yet fully realised. One example has been given for semi-pelagic trawling. A second suggestion is that of adding fins to the 'V' form board as shown by Figure 21. It is felt that these would give stability to the 'V' form board which would otherwise be lost whilst it is not in contact with the seabed. Such an addition would be inexpensive, have no contrary effect for ground trawling and may, with correct rigging, fish suitably in the middle water layers. These

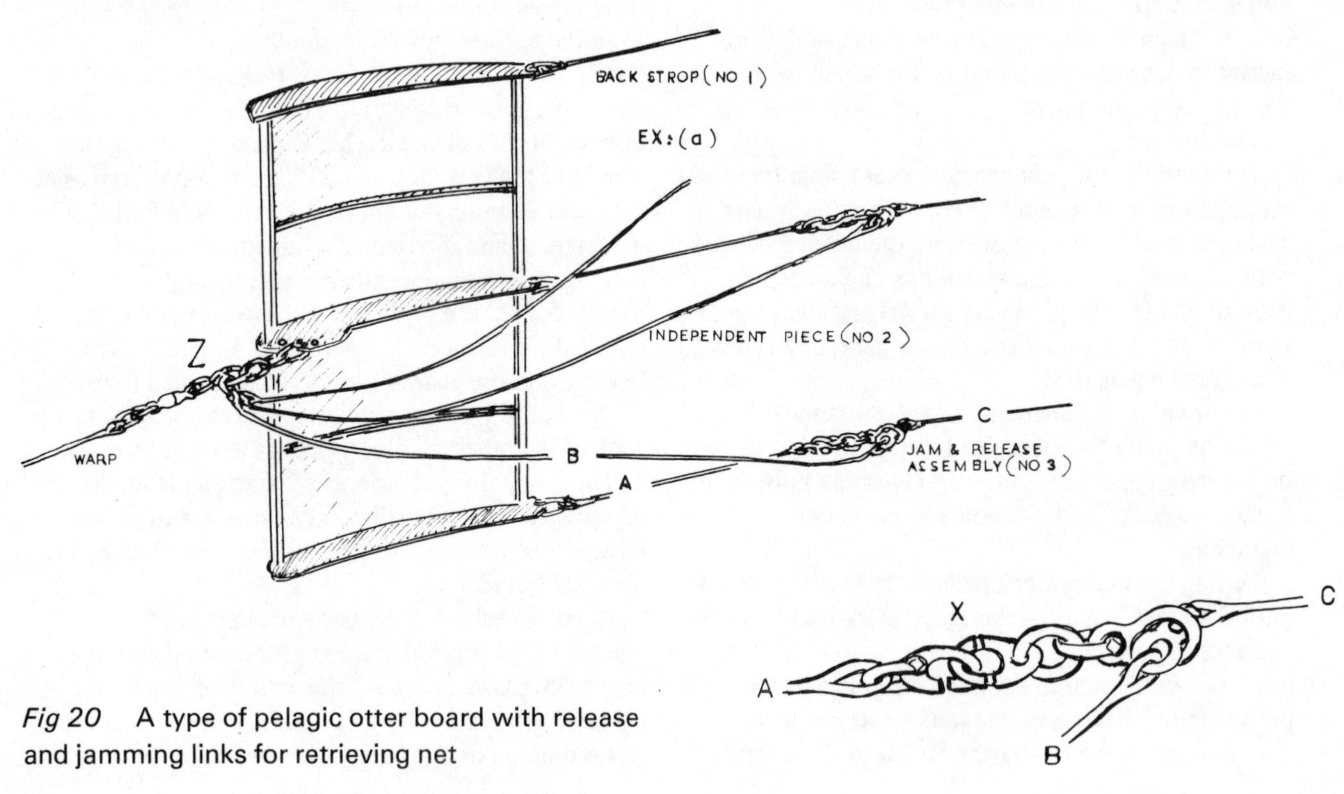

Fig 20 A type of pelagic otter board with release and jamming links for retrieving net

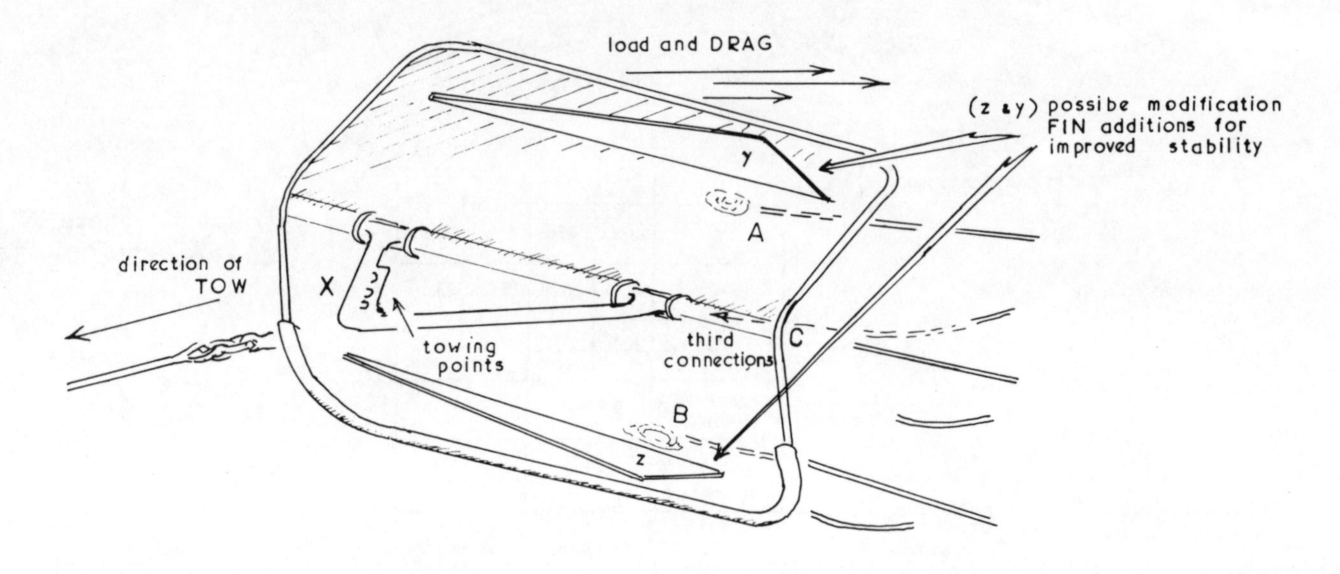

Fig 21 Vee form otter boards (for dual purpose trawls)

fins, assumed by z and y on the drawing would not need to be very deep, and could be rounded, so as not to be in any way an encumbrance.

Japanese Pelagic and Semi-Pelagic Otter Boards

Figures 22 and 23 show two types of otter board in use from Japanese vessels for pelagic and semi-pelagic fisheries. It will be seen that the shape, whilst concave, similar to the hydrofoil otter board, is somewhat different, and thus assumes a more substantial appearance.

In each case, in common with modern otter board development, there are three main towing positions for the attachment of the warp, whilst at the rear of the board there are four back strop connecting positions.

The smaller sketch on Figure 22 shows the angle of attack in degrees, and relates the towing connections 1, 2 and 3 with the appropriate back strop connections for assuming assessed angles of attack.

The small chart presents the various angles achieved with the different settings, and the weights of these otter boards (in air and water) are given at the foot of the chart.

A Simple Release and Connection Assembly Developed in Canada

For the smaller type of semi-pelagic trawl where two bridles of extensive length are required to permit good vertical lift to the net, a simple jamming device and retrieving assembly was thought up in Canada. This is shown in detail by Figure 39, and as the weight is carried mainly along the groundrope A with the lesser weight along the bridle B to the headline, the jamming was effected at X in the normal fashion. When the otter boards arrive at the gallows they are released with the G links at C and thereafter, as the weight was taken along the warp D the independent piece E begins to haul in B (the bridle to the headline) until the load is released from the jamming link at X, and bridle A is then free to haul through X.

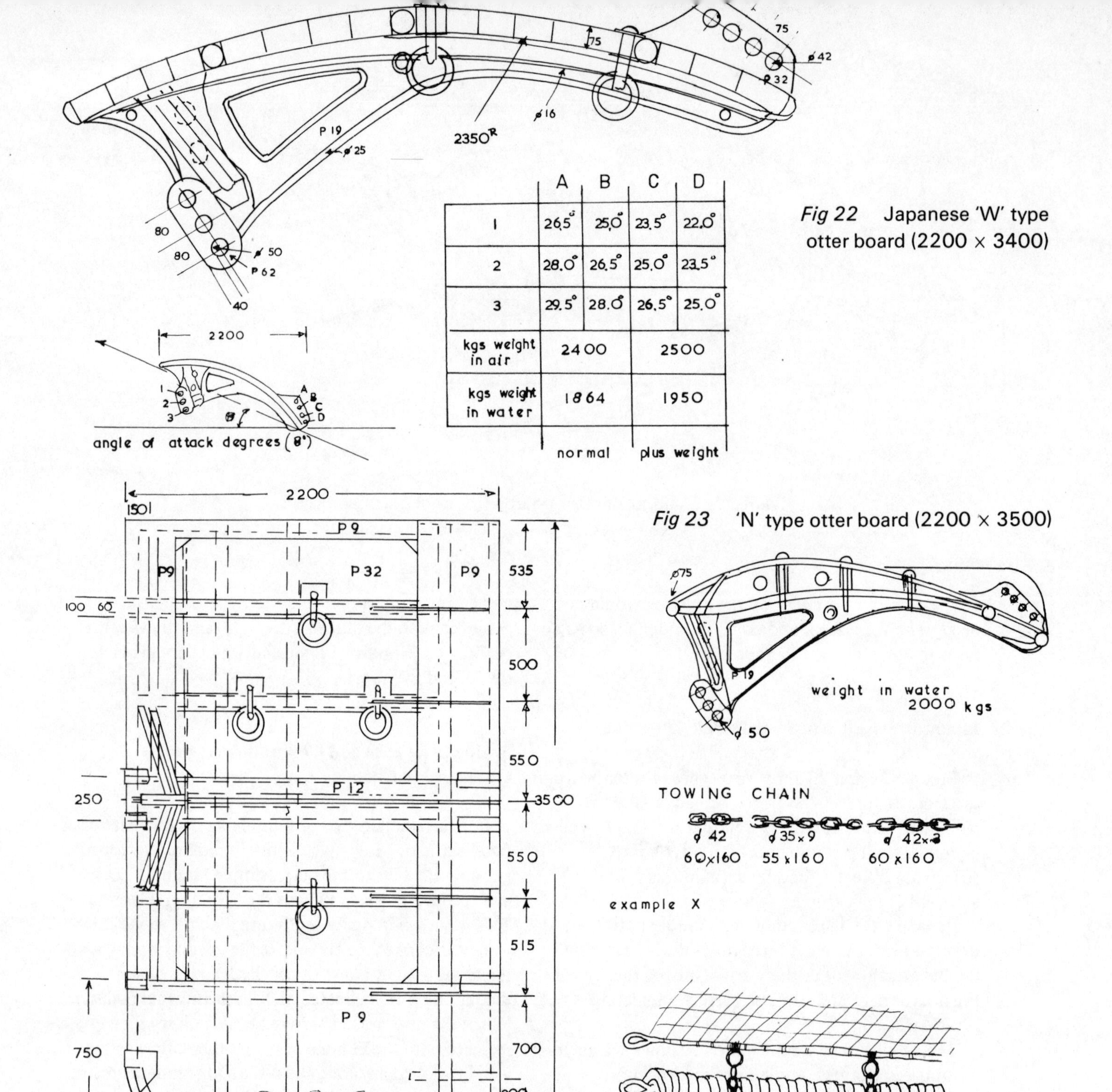

	A	B	C	D
1	26.5°	25.0°	23.5°	22.0°
2	28.0°	26.5°	25.0°	23.5°
3	29.5°	28.0°	26.5°	25.0°
kgs weight in air	2400		2500	
kgs weight in water	1864		1950	
	normal		plus weight	

Fig 22 Japanese 'W' type otter board (2200 × 3400)

Fig 23 'N' type otter board (2200 × 3500)

5. Notes on Rigging Procedures for Pelagic Trawl Nets

Pelagic Nets Strengthened with Support Ropes

Pelagic trawls are, in comparison to ground trawls, very large nets. And whilst the mouth of a pelagic trawl is usually made with large meshes, the body netting is graduated as it narrows, with decreasing mesh size of the lighter ranges of nylon twines. In short, most designs are voluminous, comparatively weak and susceptible to sudden load factors. To counteract this, pelagic trawls have to be strengthened far more than are ordinary trawls, and it is quite normal, in addition to the headline, footrope and side ropes, to have large nets, which are manufactured for fishing mid-water, encased in a framework of support ropes, along which the stresses are carried and transferred to the main towing ropes.

Naturally, the grading of the support ropes is determined by the particular design of the net, that is the pattern, whether it is a two-, four- or multi-seam net. For example, Figure 24 shows a medium size type of four-seam pelagic trawl being towed by one vessel, with the main support (or strengthening) ropes clearly emphasised. The principal support ropes are those attached along the seams (or lastridges) of a net, for it is along these important ropes that any proportionate slack netting requirement must be securely held and carried (see Figure 26, example (c)). The 'seaming' ropes, as they may be termed, are normally

Fig 24 The main support ropes on a single boat four seam pelagic trawl net

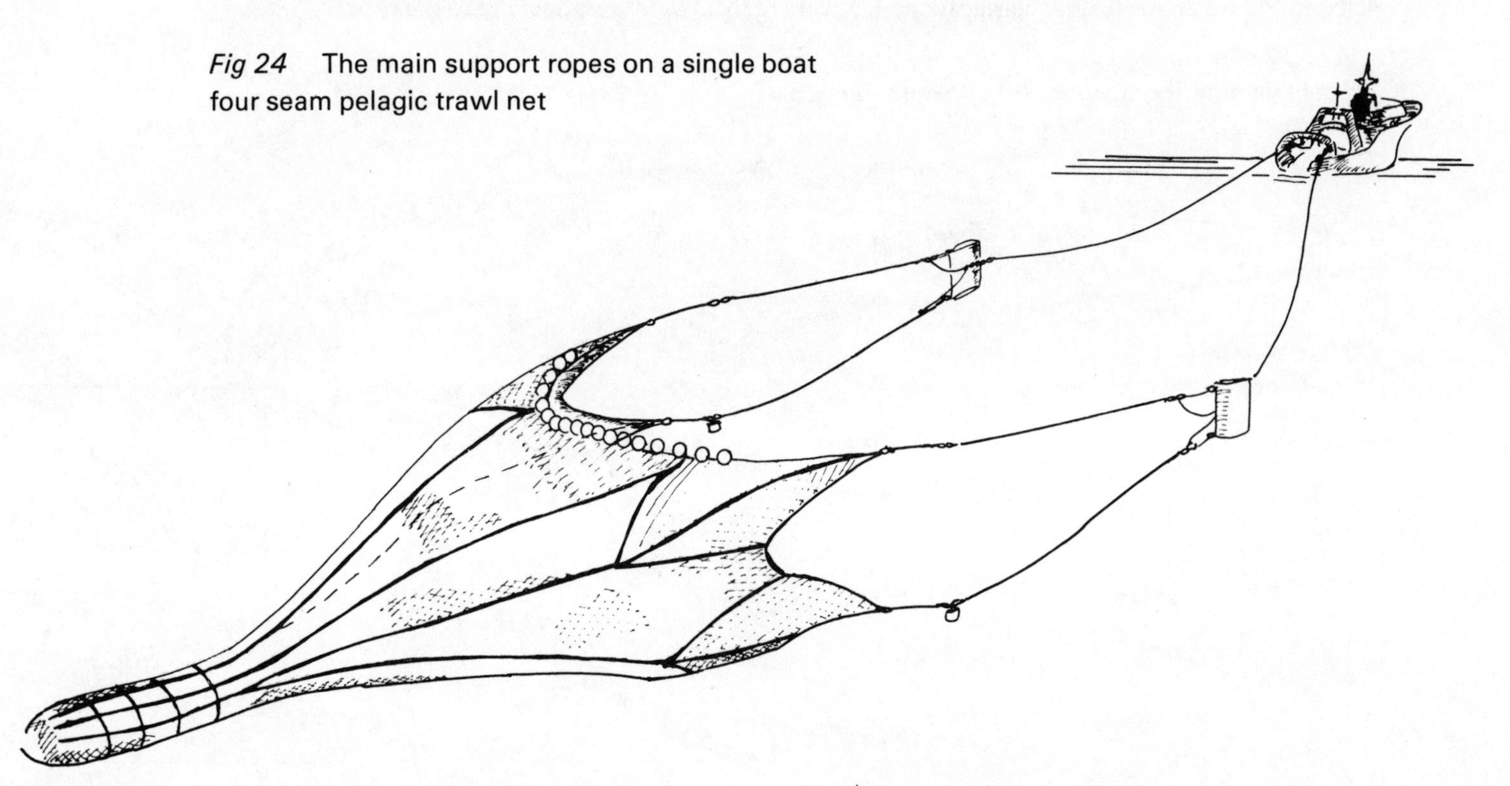

heavier than the other support ropes, for they have to be strong enough to carry a large percentage of the load directly to the towing bridles.

Similarly, all pelagic trawls are additionally strengthened from the top and lower 'quarters', – that is, where the wings meet the body netting at the 'bosoms', to the codend, and where proportionate allowances of slack netting have been made on seaming lines, so comparable ratios would be necessary on the lateral lines. It should be mentioned at this stage that there may be more than four lateral lines, *ie* one from each 'quarter' for with very large nets several lines may be necessary to give strength laterally between the headline and the codend.

Also, as shown, further rope support is given diagonally from the 'quarters' or wing joins, to the seams. This is perhaps more clearly illustrated on Figure 25 with the enlargement (a). Thereafter the codend and sometimes the lower part of the body netting is strengthened with numerous cross ropes. Figure 25 also shows the release assemblies, and thus can be related to Figure 20 for further clarification.

Reinforced Netting – Double Corners, Wing and Bosom Meshes

In common with all kinds of nets the main working edges or parts of the netting which attach to the main ropes are reinforced with meshes of double thickness, or heavier twine. Figure 26 gives an example (b) of how the side panel trawl wings might appear on a small pelagic net. A similar picture, although enlarged, might be applied to other leading edges of all trawls made for fishing the middle water layers.

With original standard ground trawls the curve of the headline and footrope was shallow, with the bosom meshes hung to a 50% to 70% ratio of the stretched length, the top wing more or less roped at stretched length along the 'halfer', whilst the lower wing would be set in 10% to 25% but still diagonally along a halfer. With increased trawl net sizes, and the development of pelagic net patterns, more sweeping curves of wider angles were desired, thus the forward edge netting was made to adopt better curved formation for attaching to the ropes.

Fig 25 A pelagic gear for one vessel (four seam design)

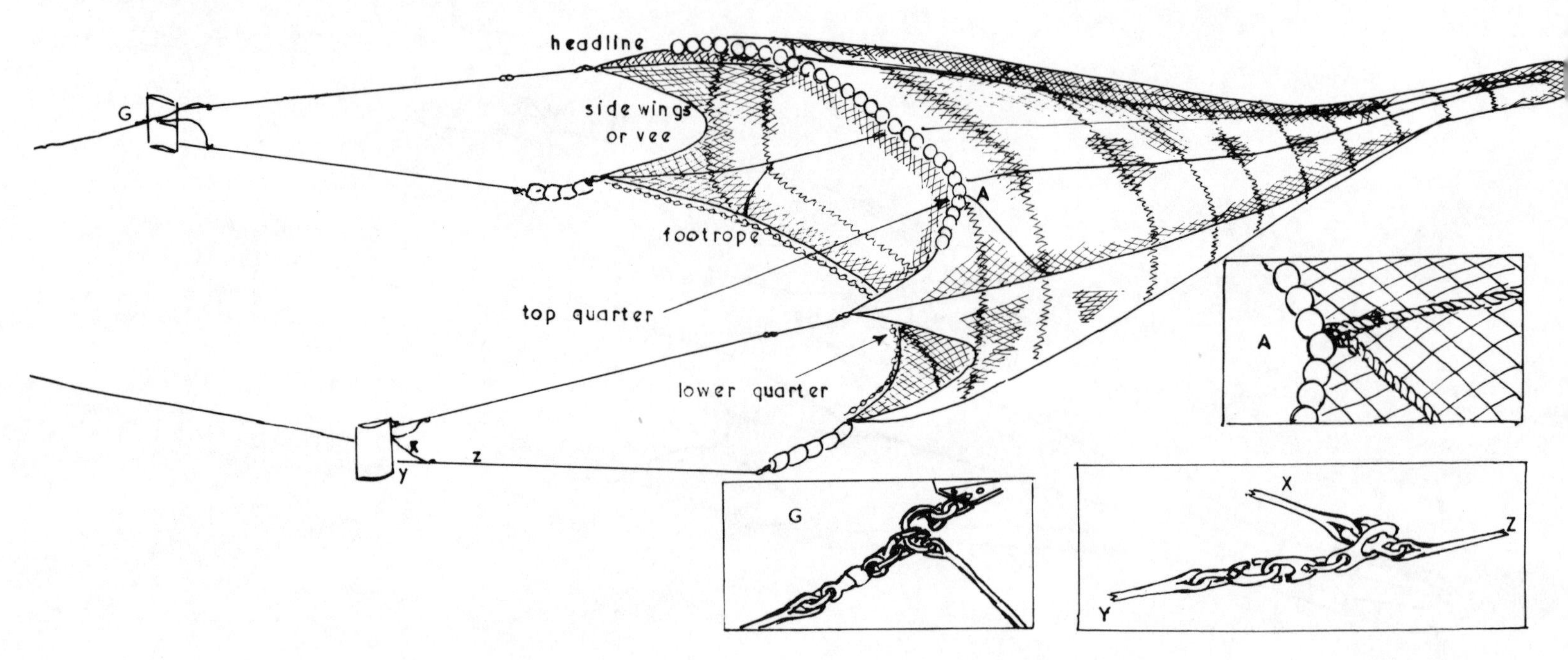

Figure 26, example (b), gives a very simple indication of achieving an improved curve, with X=4 bosom meshes, Y=4 two bar meshes, and Z along the halfer. The example featured would of course be the side panel and wings of a very small net, but the basic objective has been shown. To enlarge on this, a net may have 50 or maybe 100 bosom meshes, and extensive wings, so a more accentuated curve might be required; thus the wings may start with one bar meshes then go to two bar meshes and perhaps to three or four bar meshes, to achieve the desired dimension. And in fact there are variations of mesh formation for different patterns, with the edges and main rows made of heavy or double thickness twines.

To demonstrate these points further, if reference is made to Figure 27 it will be clearly noted from this drawing, how the meshes are graduated, from V (the bosom meshes) through W and X to Y (the diagonal wing meshes). The netting is shown opened to its widest capacity, and the curve would be far less sweeping if the netting was flowing by half, as illustrated by example Z.

The even flow of the netting of pelagic gears, with sweeping curves to the ropes is all important, and of course many factors have to be considered when roping the mouth of the trawl. One is the percentage of slack netting to be given (if any) along the seaming lines; this then has to be co-ordinated with an

Fig 26 A and B illustrating strengthening on side panel trawl wings of a small four seam net and C showing slack netting rigged to rope

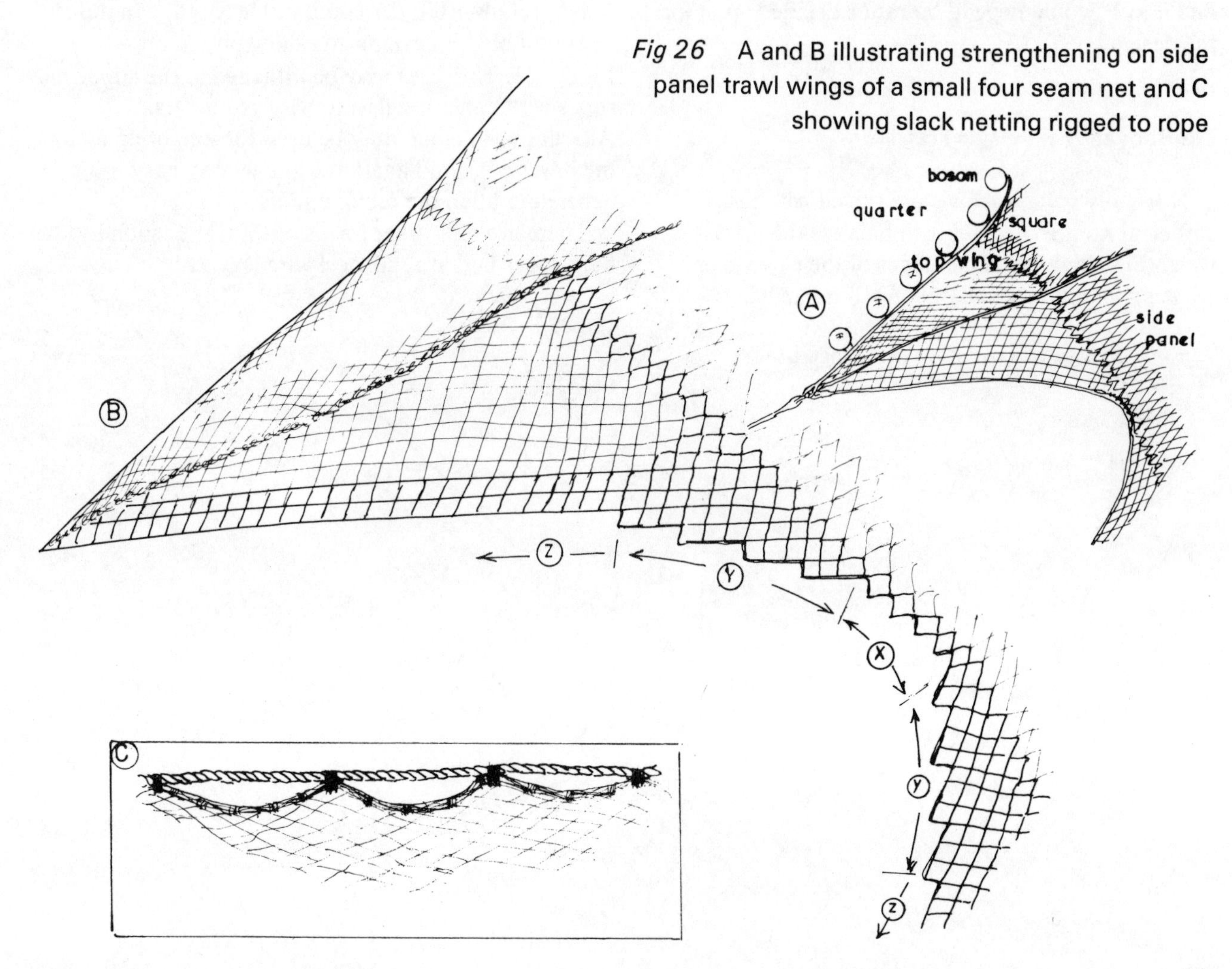

understanding of the hanging ratios of the netting around the mouth of the trawl.

Mesh dimension and percentage hanging ratios are covered in my book *How to Make and Set Nets* (also published by Fishing News Books Ltd) and from the information therein it would be possible to calculate the correct percentage ratios of slack netting to the seaming lines, in relation to those of the main ropes *ie* headline, footrope, sidelines.

Basically, however, the hanging ratios on the main ropes, depending on the mesh formation and the shape of the netting required, are mathematical, as shown by Figure 27 with netting b = to rope length W, netting c = to rope length X and the lines a to V and d to Y would depend on ratios decided upon for the design.

The Buoyancy of Pelagic Trawls

Normally pelagic trawls are rigged with just sufficient numbers of floats to balance the precise weighting along the lower edges of the rig to give what is sometimes termed 'neutral buoyancy'.

Rigging practices do vary and of course there are many special types of shearing floats for achieving maximum 'uplift' through the water.

The main objective is to achieve the best vertical opening without the trawl tending to sink or lift or to adversely affect the spread. Such information regarding numbers of floats and loadings of particular nets is readily available for specific specifications.

One of the main types of weighting along the lower edge of the mouth of the trawl, which may be regarded as a footrope, consists of compact rubber discs cut from old tyres. These vary in diameter and consequently, depending on the size of discs used, the weight relationship per comparable length can differ considerably. An example is shown by X on Figure 23. The discs may be attached to the support ropes with chain toggles, or wire grummets. Alternatively, chain may be used for weighting along the lower edge and again the gauge may vary depending upon the requirements.

There are also other forms of footropes such as fibre ropes laid around steel wire legs, *etc.*

Fig 27 How the meshes are graduated

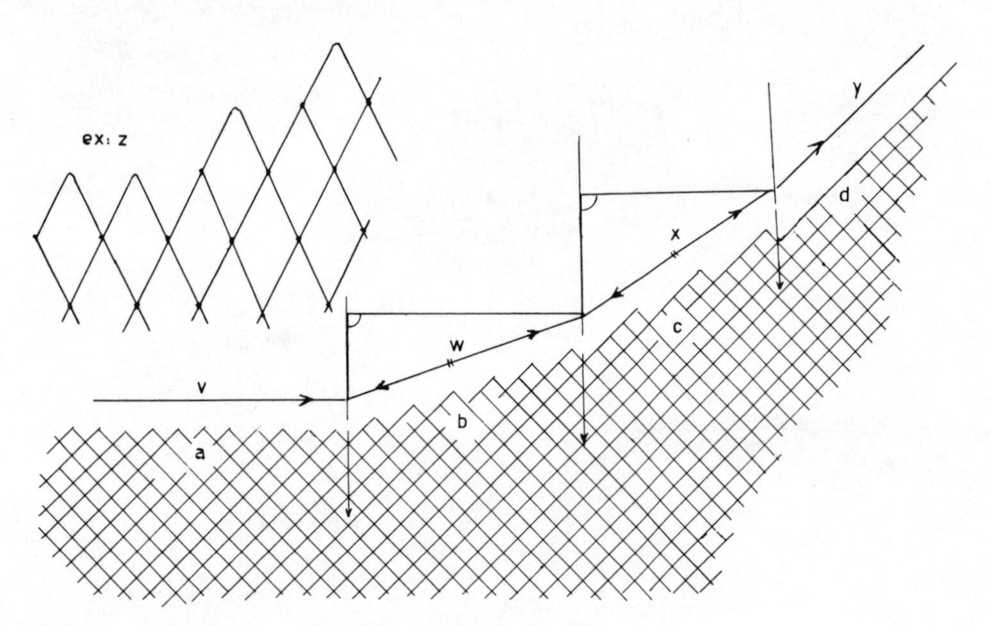

6. Semi-pelagic and Single Vessel Dual Purpose Trawls

One of the first successful semi-pelagic trawls was that featuring the otter boards worked singly from the lower spreading wires, with the top spreading wires being an extension of, or connected to, the warps. Additional towing wires joined the otter boards to the warps at similar points. Some early single vessel mid-water trawls, such as the Icelandic Breidfjord trawl, were rigged in such a fashion, but in view of the progress made with pelagic trawl operation in the last decade or so, it would seem correct to classify this assembly under the heading 'semi-pelagic gear'.

At this juncture it may be beneficial to examine the qualifications for rating some rigs as semi-pelagic and others as pelagic.

The objective of designing pelagic gears is quite apparent, there being species available at certain periods for capture in the area of mid-water to the surface. The reasons for semi-pelagic gears is not always so apparent however. Two important considerations are (1) fish located just above the seabed, and (2) uneven bottom conditions causing frequent net damage. Both these factors can be, and are, met by the use of ordinary ground gears. For instance, standard type otter trawls have been used off the rocky coastline of north-east Canada, and, it is suspected, in other regions, fishing in a semi-pelagic manner simply by using increased speed.

Alternatively, ground trawls have been designed with many improved features to give greater 'spread' and much better lift, whilst still firmly in contact with the seabed. These are normally referred to as 'high opening trawls' and many of the different design developments and specifications of new types of ground trawls are covered in the book *Modern Deep Sea Trawling Gears* (also published by Fishing News Books Ltd).

Note: It should be noted that the bouyancy of a float does not decrease with depth of water.

There is, however, a precise difference between high opening nets designed for ground contact, and gears made to fish as semi-pelagic trawls. The semi-pelagic trawl gear should be assembled so that the otter boards are in contact with the seabed, whilst the trawl net is rigged correctly with floats, weights and bridle lengths to fish a metre or so off the bottom.

There are two main methods of achieving these results. The first is as briefly outlined at the beginning of this chapter, and the second is a three-bridle assembly, correctly proportioned, with oval, 'V' form or rectangular otter boards, joined to a high opening net with high flotation values.

To examine these two principles in turn, Figure 28 shows a four-seam semi-pelagic trawl, and rigging assembly. F illustrates the towing warps, D the connection between the otter boards and the warp. E the top spreading wire (or warp extension). A represents the otter board, and C indicates the lower leg between the back strop and lower wing end. This is further exemplified by Figure 29 which has additional identification letters: *viz* h the headline and g the footrope. The sketch Figure 29 is also meant to give the impression of the otter board being in ground contact, whilst the net with 'neutral buoyancy plus' is floating so that the wings at X are well off the seabed, as is the main body netting as illustrated by b.

A further drawing, Figure 30, represents a side trawler hauling such a gear, and Figure 31 gives a specification of a particularly successful net and assembly.

The second form of semi-pelagic operation, a high opening trawl net, rigged to fish off the seabed whilst the otter boards maintain ground contact, is primarily illustrated by Figure 32 which shows oval otter boards lightly dredging over the seabed, long backstrops, single bridles, three extensive spreading wires, adjusted to make the trawl net fish off the

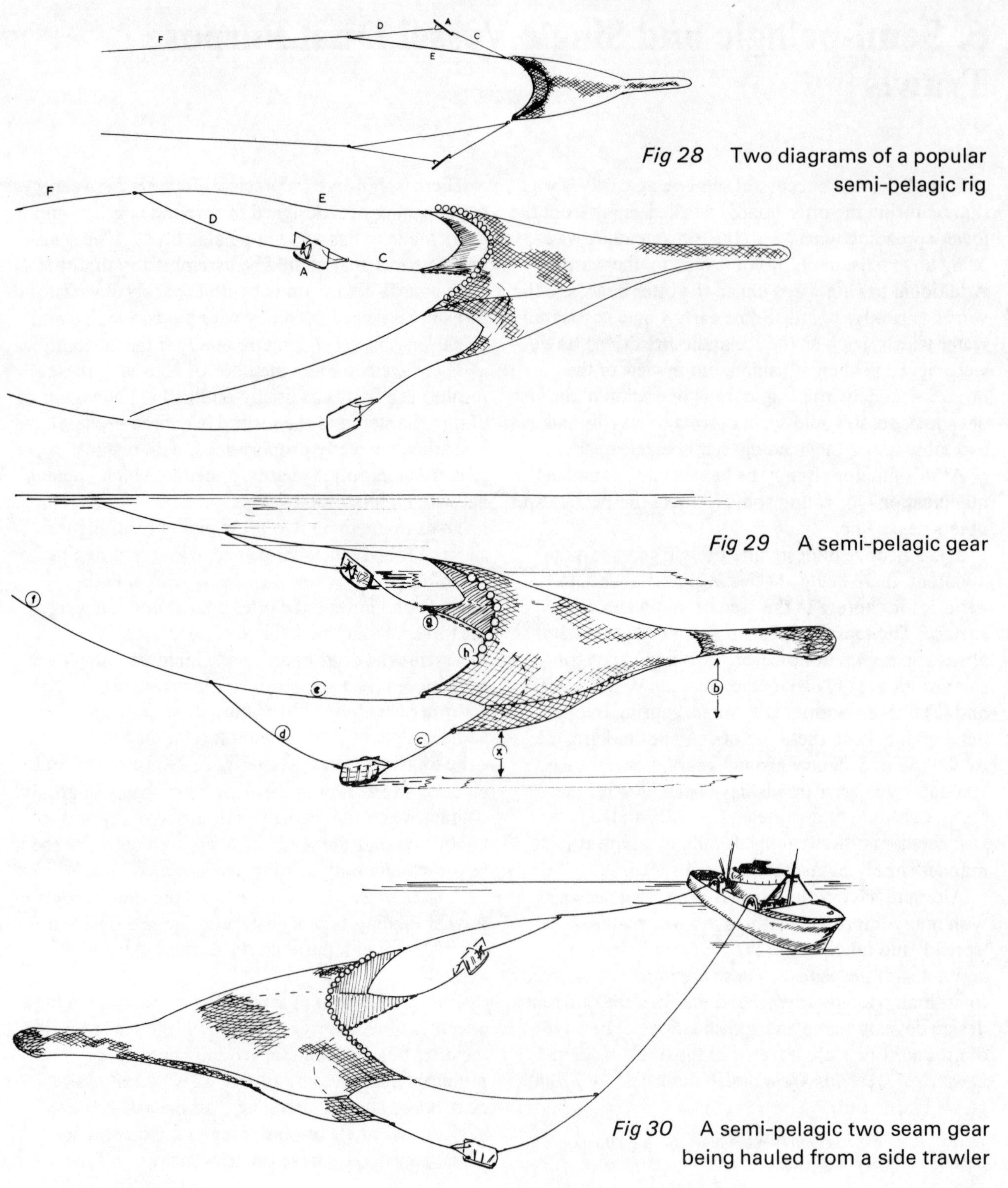

Fig 28 Two diagrams of a popular semi-pelagic rig

Fig 29 A semi-pelagic gear

Fig 30 A semi-pelagic two seam gear being hauled from a side trawler

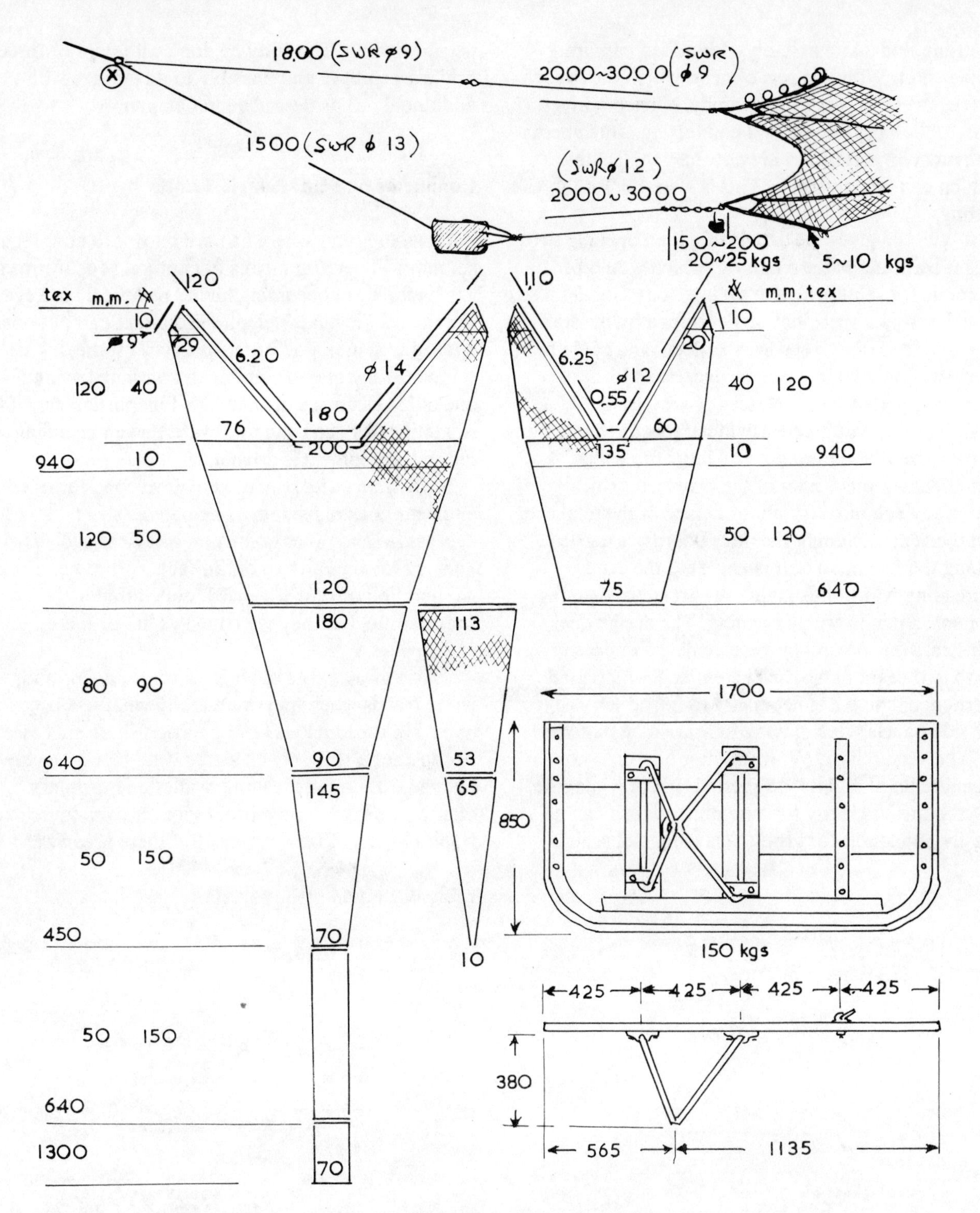

Fig 31 A four seam pelagic trawl for a trawler with 150/200hp engine

bottom, and a kite assembly for added capturing dimension. With this sort of gear, which the French have developed very successfully, efficient swivel assemblies and release and connecting attachments are most important, as are correct wire lengths, which can be adjustable. This is also true for the kite fittings.

Figure 33 gives the details of a high opening two-seam trawl net pattern and rig, which would be suitable for semi-pelagic trawling from a medium size trawler. It is a specification developed by the staff of the Marine Laboratory in Aberdeen, and could be proportionately increased or decreased to accommodate various classes of vessels.

This specification clearly illustrates the tremendous cover given by the square and large top wings. A good deal of the weight of the trawl net, which normally accounts for about a third of the total drag on the gear, including the otter boards, is carried along the two main seams and onto the central spreading wires, thus giving the greatest advantage for spread and vertical opening. The design does illustrate the maximum to be achieved dimensionally with two-seam patterns. Thereafter for increased vertical opening it is necessary to introduce wedges or side panels, thus moving to four-seam patterns.

The versatility of 'V' form otter boards is now being realised and it is suggested that they may be used quite efficiently for assemblies which can be easily adjustable for ground trawling and semi-pelagic fishing. This may be done with two- or three-bridle assemblies and possibly to advantage with additional fins as described in Chapter 4.

Comments on Semi-Pelagic Trawls

It is a fact that where fish are known to congregate so man will develop means of capture. Most forms of fish capture are becoming highly technical, and even simple 'walkie talkie' telephone contact can put open boat fishing into a different sphere. In some degree a skipper today may be unable to continue fishing if one of his electronic fishing aids fail, particularly if he is 'aiming' for shoals in mid-water, when his reliance on such equipment is crucial.

Early mid-water trawls were developed for river and near coastal fishing, using canvas sheet depressors and other home made devices such as kites, paravanes and so on. In fact, a varied conglomeration of gear was tried, quite often successfully, but they were always 'hit or miss' endeavours.

This was also true of claimed triumphs for mid-water trawls when they were really what has been named in this book as semi-pelagic fishing successes. For instance, trawls such as the wing trawl, heavily buoyed with long spreading bridles and ordinary otter boards, were used profitably for fishing the mid-water regions of many areas. But these areas were

Fig 32 A semi-pelagic assembly with oval otter boards

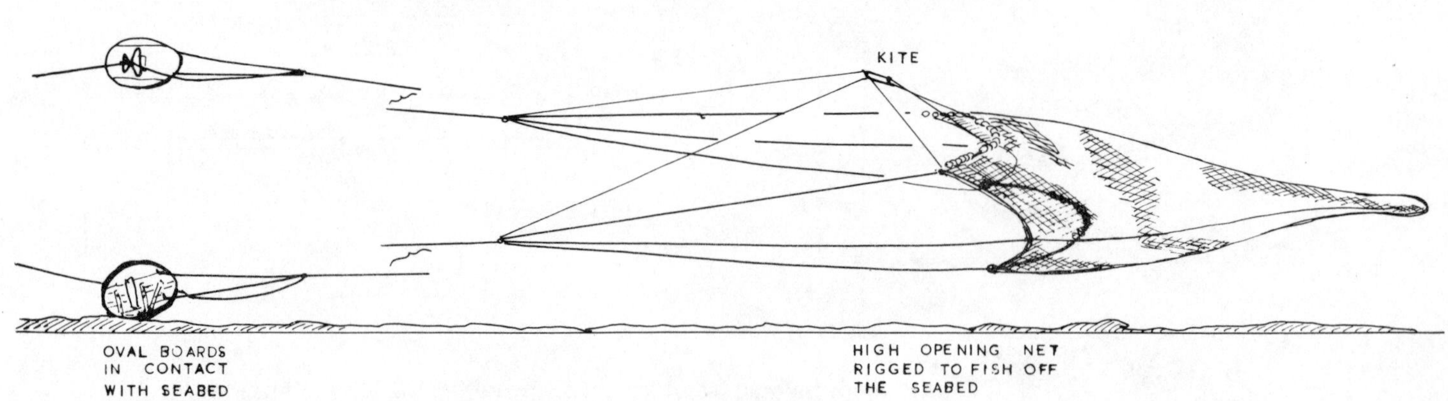

usually comparatively shallow, and the wing trawl correctly rigged would achieve tremendous 'gape', so that whilst the boards were often bouncing on and off the bottom the net itself was automatically covering wide areas between the surface and the seabed.

The precision required in deep sea fisheries nowadays is quite remarkable, however, and pelagic fishing at the different water levels is a skill to be learned.

Figure 34 is included to give the reader an impression of several vessels fishing 'off the bottom' and the scene could be far more crowded when good 'fish spots' are located.

Figure 35 is the specification of a Japanese 'high' opening ground trawl or semi-pelagic gear, depending on the fully rigged assembly used. It was a design introduced into this country almost ten years ago. It was tried as a high opening ground trawl with

Fig 33 A high opening two seam trawl net pattern and rig suitable for semi-pelagic fishing

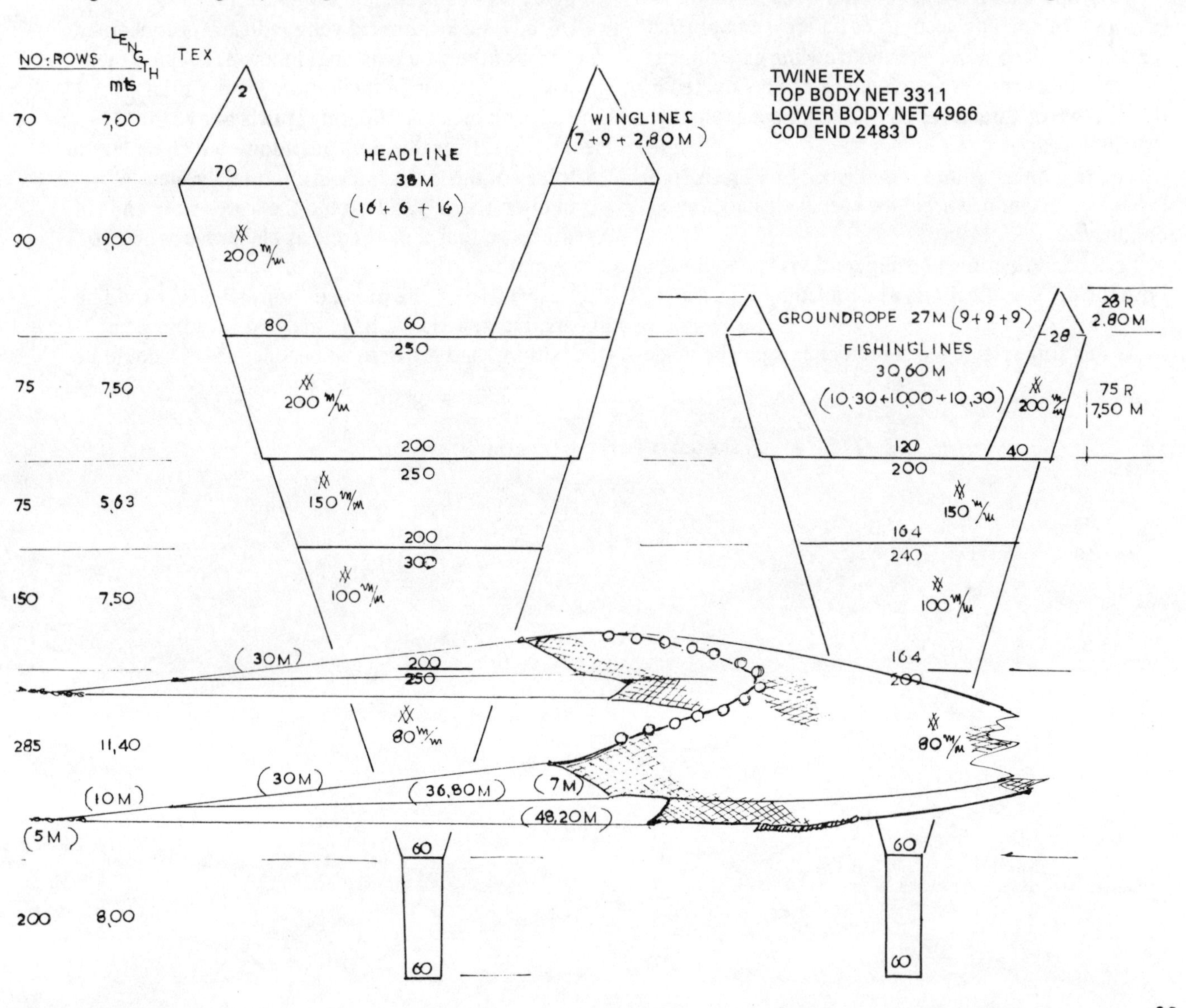

ordinary otter boards on a medium class side trawler without any real enthusiasm and was soon discarded.

Nevertheless as sometimes happens in fishing experience, in time to come a similar pattern may be tried again and found to be what every one has been searching for. As this book deals only with pelagic and semi-pelagic gears, the heavy bobbin rig which can be used in conjunction with this net pattern is not shown, but it should be mentioned that fully assembled as recommended for high opening ground trawling and with the correct otter boards (Figure 23), it would seem to be an ideal equipment for the very deep water ground trawling experiments for new edible species which have been conducted in the last two or three years by government sponsored British vessels.

The rigging assembly is described by Figure 36. It is one which can be varied for specific fishing conditions.

The codend outlined by Figure 37 is suitable for both the net specifications as detailed by Figures 19 and 35.

At this stage reference should once again be made to the release and connecting assemblies which are so important for hauling and shooting trawling gears. For stern trawlers the length of the back strop is to a large degree dependent on the distance from the main sheaves to the depth of the ramp, with clearance for release or connection on the main deck. The heavier the gear the more elaborate the assembly may be, particularly on large stern trawlers, which are far better equipped for handling pelagic gears. The equipment for small inshore vessels fishing with pelagic gears can often be simplified.

Usually the inshore trawling industry is concerned for most of the year's fishing time with keeping their trawls in contact with the bottom, rather than considering means of floating trawls above the seabed. To effect this they put additional chain legs in the place of the lower back strop on their steel 'V' form otter boards, and extend the lower wing ends of their nets so that they dredge as close as possible to the ground.

Nevertheless there are occasions when it would be most advantageous to have the gear skimming 'off the bottom' and it would be beneficial if this could be

Fig 34 A sketch of 'off the bottom' trawling

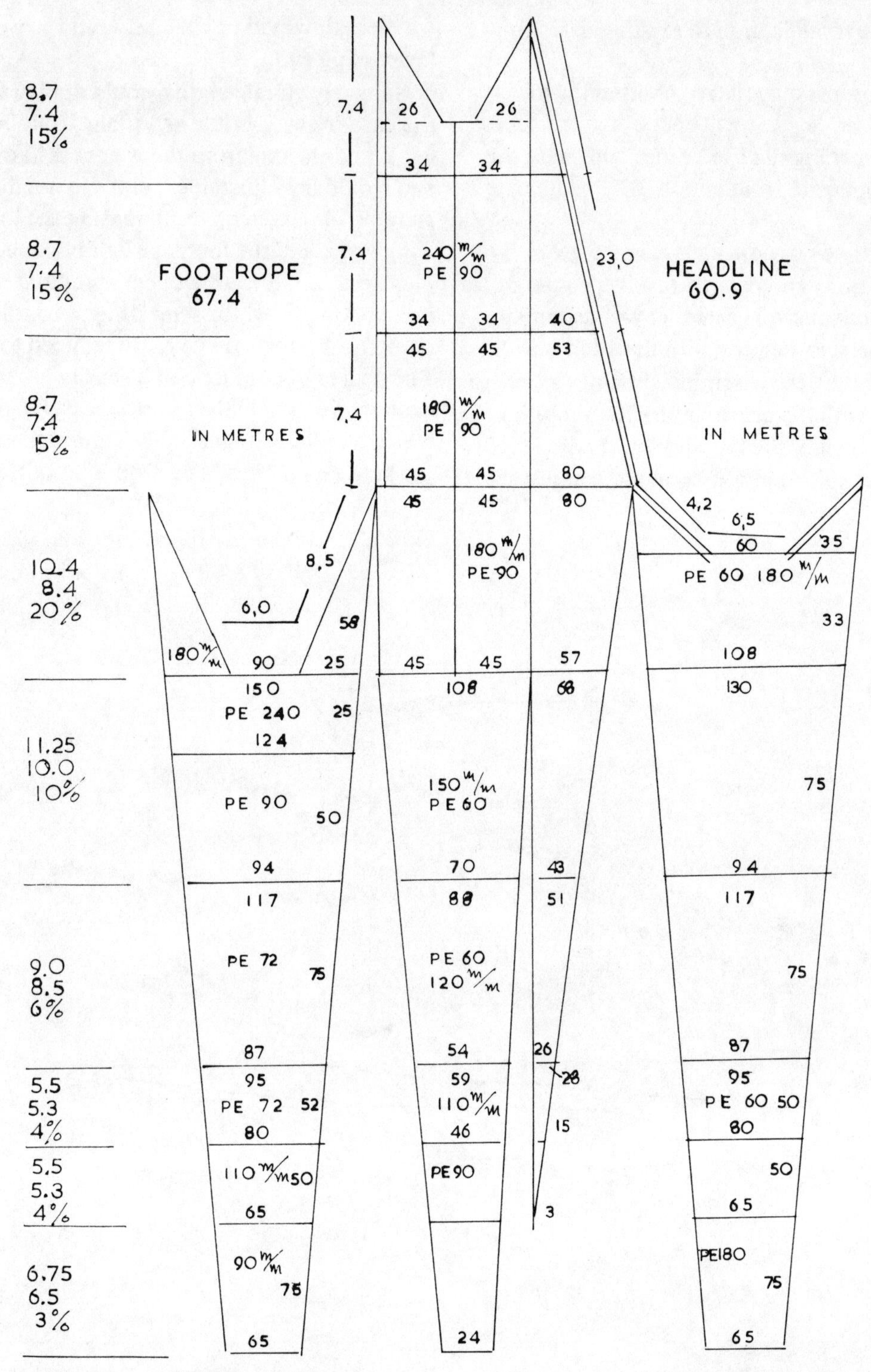

Fig 35 A Japanese 'high' opening ground trawl or semi-pelagic gear

done with simple modification to existing fishing gears.

It is known for instance that to shorten the top bridles on a two bridle gear will encourage the net to 'dig' whilst lengthening the top bridles will cause the net to lift. The opposite would apply when adjusting the lower bridles.

There are methods of adjusting existing gears for particular conditions and desired results (Figure 38). Figure 39 presents one suggested way of assembling a net for semi-pelagic fishing. With the additional fins A and B for stability, the assembly shown is a two- or three-bridle rig, with the centre bridle X–Y. With a two-bridle gear each of the top and lower wire lengths would be comparable to line Z–L and the adjustable wire length would be that between the weight and the lower wing end.

Similarly equal lengths would apply (up to Z–L) in the case of three-bridle assemblies, but both the short leg Y and the one from the weight to the lower wing end would be adjustable. With the weight made suitable for lessening or increasing the load at the wing ends, and the footrope lightly chained with a plus flotation, the adjustments required to convert a box or wing trawl for semi-pelagic conditions, and back for ground trawling, are reduced to a minimum. The degree of contact can be easily assessed from examination of (1) the otter boards, (2) the weight which should be regarded as a main factor, and (3) the light chain bites hung from the footrope.

Fig 36 Ideal equipment for deep water ground trawling experiments

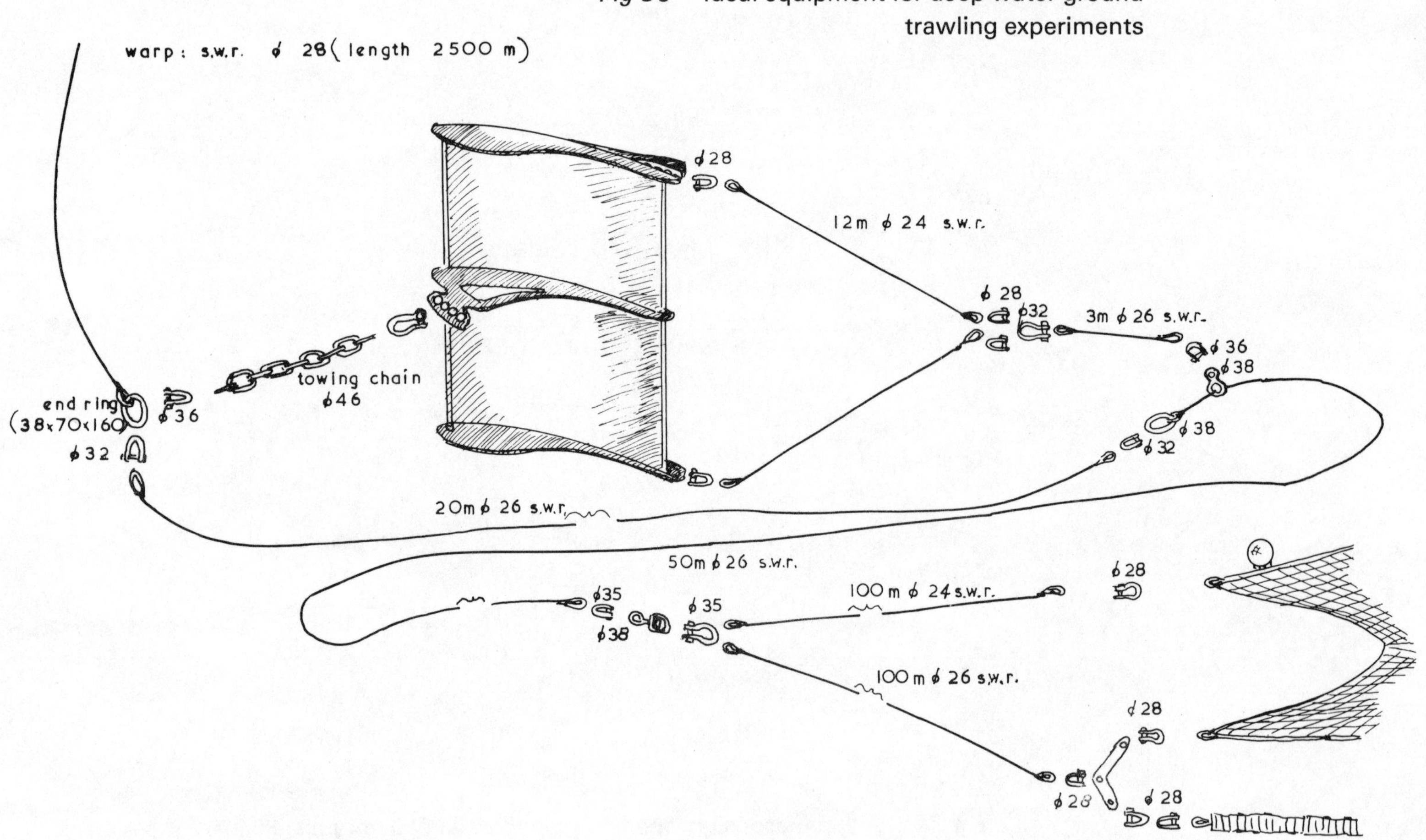

The above suggestion is one simplified method of satisfying single boat dual purpose requirements, for the occasional appearance of pelagic species.

There are, of course, specific gears in use for semi-pelagic trawling and some of these designs have been examined in this chapter. Other types are in use, for instance nets rigged with the Hong Kong paravane shearing appliance. (Figure 40.)

Fig 37 A cod end suitable for a deep sea multi-seam trawl, a high opening ground trawl or semi-pelagic gear

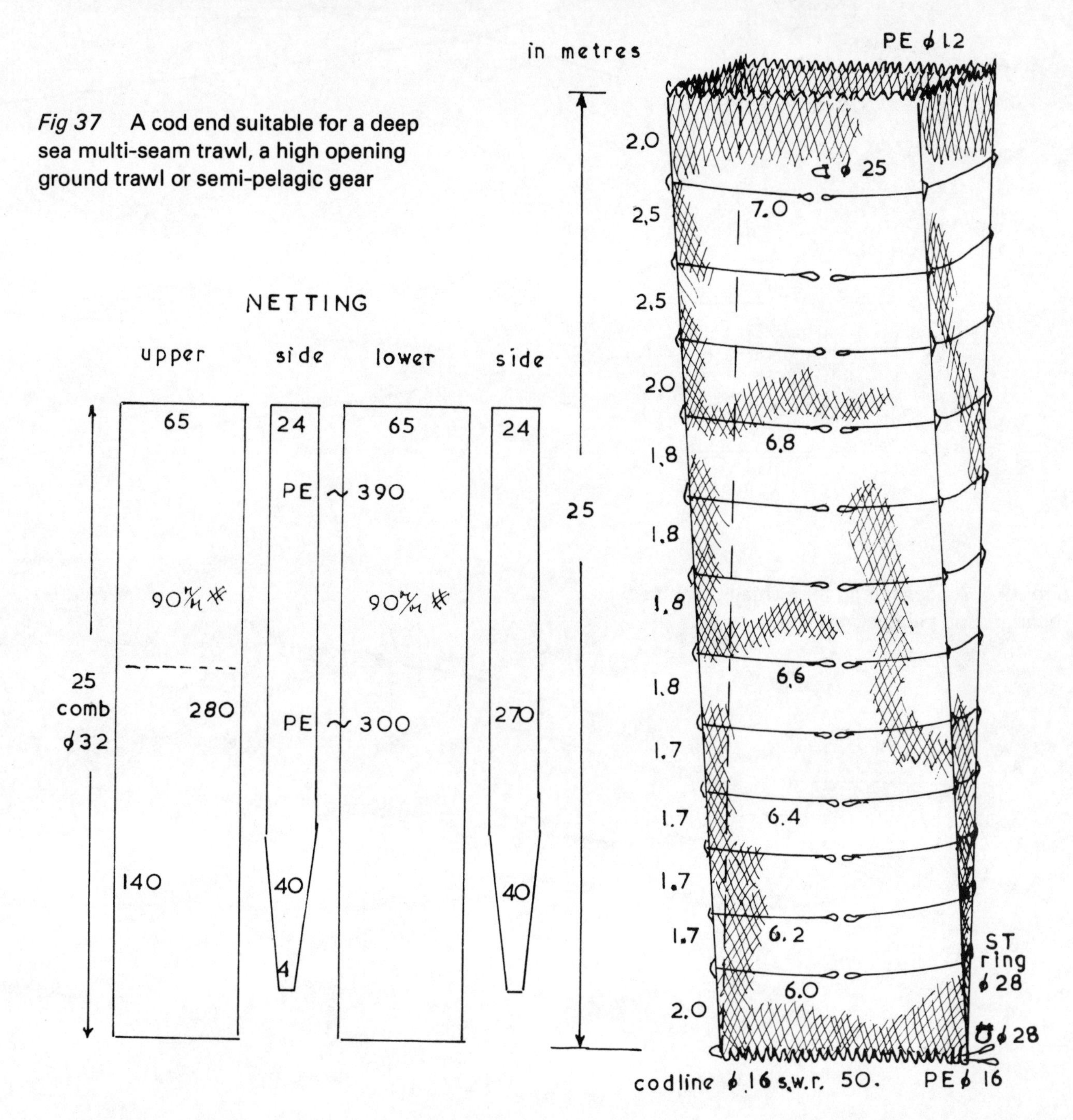

Fig 38 Canadian jamming device for smaller type semi-pelagic trawls

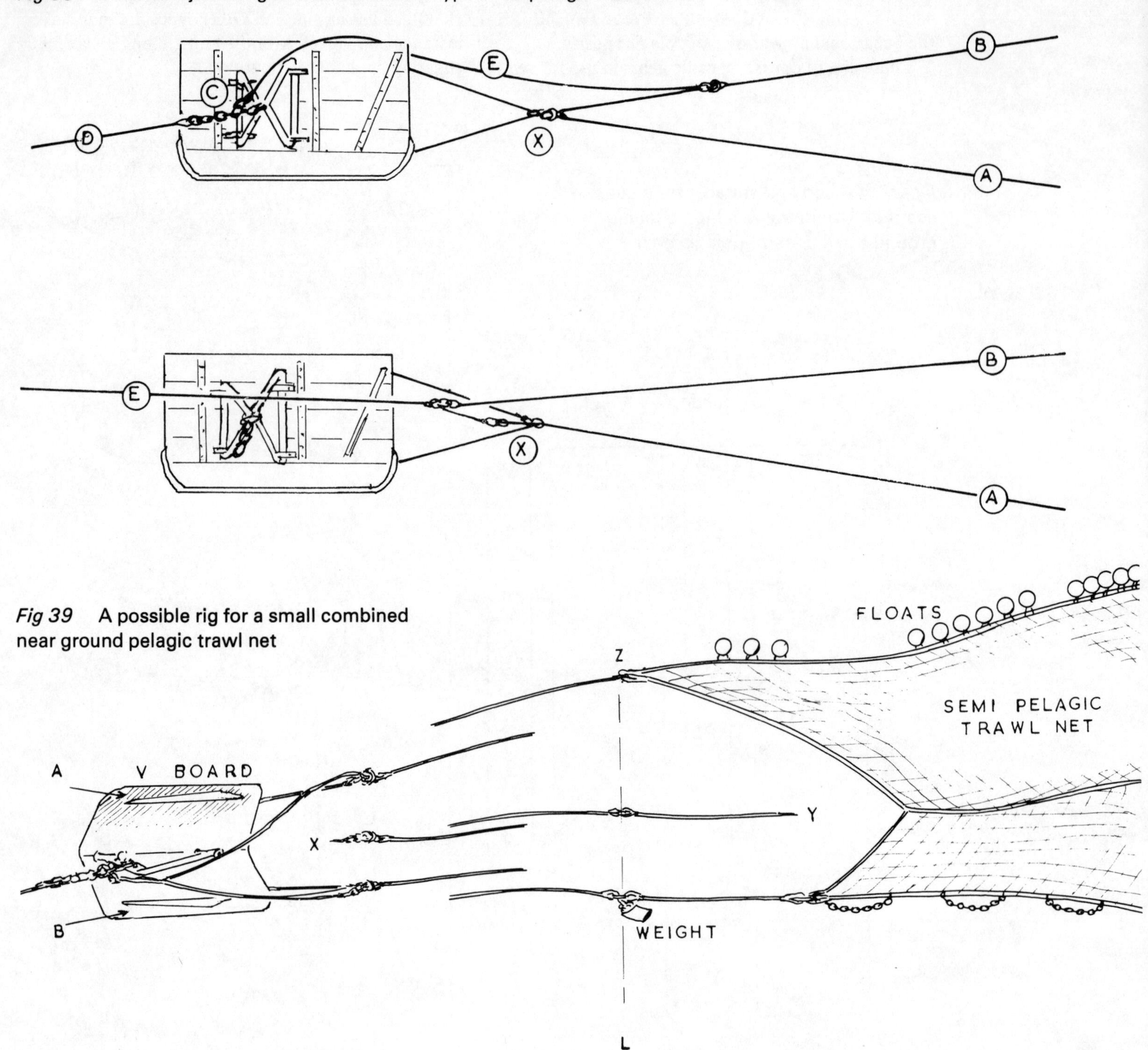

Fig 39 A possible rig for a small combined near ground pelagic trawl net

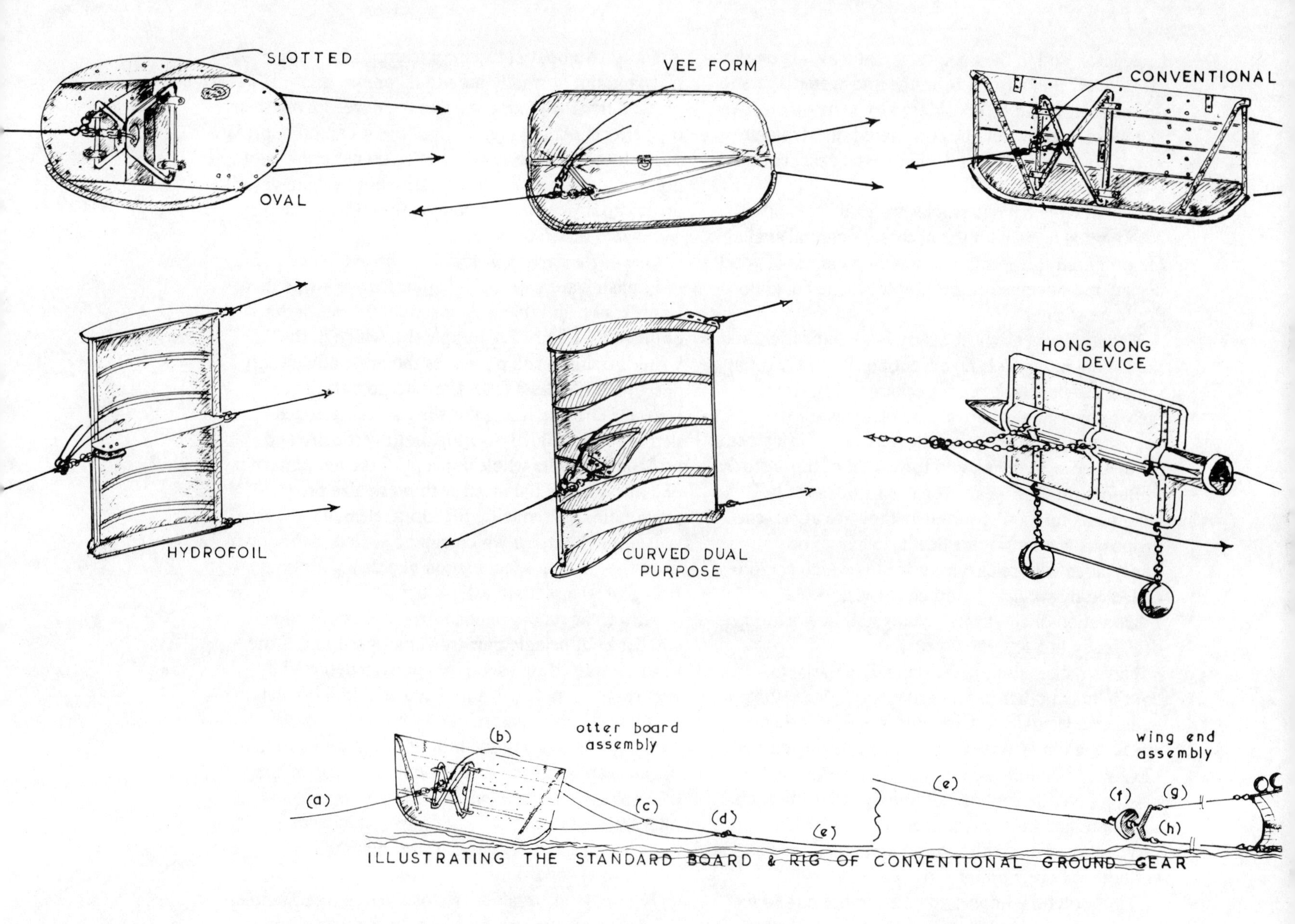

Fig 40 Different types of otter boards or shearing devices

7. Pelagic Pair Trawls and their Application

In general, and in the past, two boat trawling on the ground, off the bottom, or in the mid-water layers to the surface, could be considered to be most effective. But assessment would be considered questionable today in view of present fishing vessel design and the costs of fuels, *etc*.

Where large pelagic stocks are available and commercial firms with the necessary capital available are prepared to speculate in building vessels of good design and dimensions, profitable results could no doubt be secured.

Economics is now necessary in all industries, and the fishing industry is no exception. But there often appears to be a lack of appreciation of the main factor governing fisheries. The factor is the importance of establishing the location of fish stocks, the distances involved, and the extent of those stocks.

Before the European Northern stocks really became as rare and coveted as they are at present proposals were afoot for fleet trawling to be undertaken with releasable codends which could be collected by servicing mother vessels.

Depletion of stocks has changed that situation drastically in a few short years.

Nevertheless some of the shelves off the larger continents in other parts of the world, do feature extensive, virtually unexploited stocks of pelagic species, and it is in these regions that pelagic trawls have yet to be fully demonstrated. Pelagic pair trawling could produce resounding results in such a situation, and fleet trawling could be a feasible proposition with a third vessel unloading the pairs in turn during operational activity.

That potential applies to boats in the 60–80 feet range but pair trawling might however, be better planned in general as an operation for smaller boats.

At times when pelagic species are running in quantity two boat crews may team up for pair trawling, but normally individual crews, and particularly their skippers, like to succeed from their own efforts. And it can be said that it can often prove difficult to find individual skipper/owner crews who are really compatible for pair trawling. Where vessels are company owned, of course, the situation is somewhat different.

One of the early pair trawls to be used successfully in English waters for pelagic pair trawling was the Essex trawl, and this was based on the net pattern outlined by Figure 7, example (b), which in the writer's opinion still possesses the most advantageous design features as a four-seam net to date. The outward shaping along the wing seaming edges permits wide angled opening, both vertically and horizontally. The whole concept of the net pattern is one (if correctly balanced with mesh size proportionates) of reducing the 'drag' element.

It is a net pattern which can be scaled up or down from the existing specification necessary for boats from 20/30 h.p. up to 300/400 h.p.

Also to be recommended, particularly for the smaller boat pelagic pair trawling operation, is the more simplified two-seam net pattern depicted by Figure 7, example (a), and illustrated in more detail by Figure 5. This design allows a certain degree of versatility in shaping the mouth of the trawl, in that the cut can be accentuated in one direction or the other to give marginal increased vertical opening, or alternatively wider horizontal gape, whichever is regarded as most beneficial for anticipated operational conditions.

Figure 41 exemplifies two of the rigging principles in use for pelagic pair trawling. In the first drawing, example (a) shows a full trawl rigged with the 'twin warp method' whilst the second sketch, example (b),

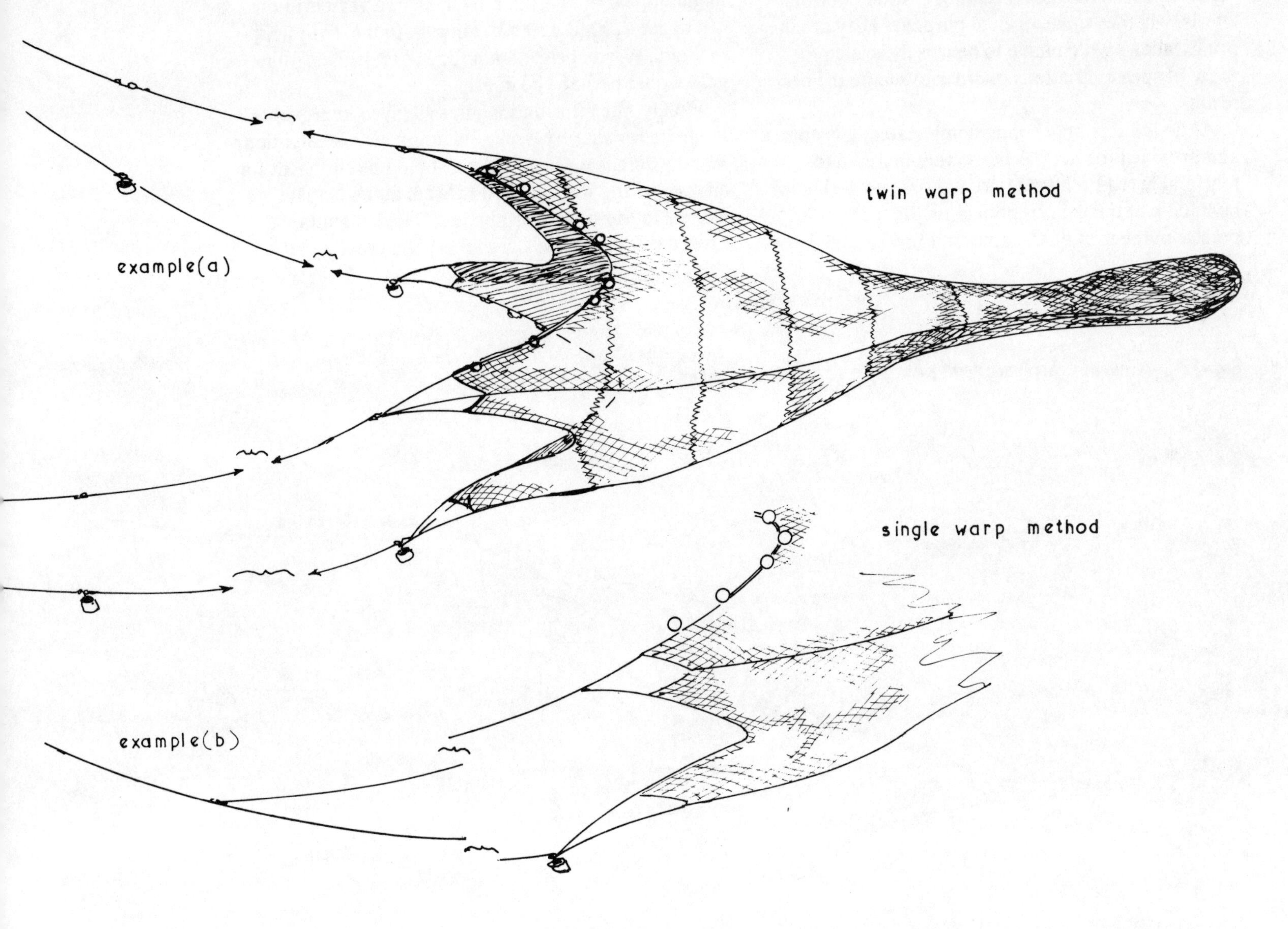

Fig 41 A two boat pelagic trawl

represents part of a trawl mouth assembled to the 'single warp method'.

As has been stressed continually throughout this book, that properly balanced weighting and flotation values are vital for the efficient operation of pelagic trawls, whether single or dual purpose. They are the prime factors which have to be closely related to those of speed variations in order to secure the best results.

Although this may sound complicated, it is not so, and provided the net has been manufactured to reasonable requirements (size to horse power) the balancing of the gear or perfecting the trim, *ie* precise number of floats, amount of weight, and correct speed requirements, become factors to be measured through empirical stages rather than to be given out of hand.

Figure 42 shows a two-boat pelagic trawl as it might appear in operation from fishing vessels in the 40′ to 50′–100 to 150 h.p. range with the 'twin warp method'. A net specification for 80′ to 100′ headline gear would perhaps be in use.

Finally, the information already given in previous chapters for single vessels, including net specifications and the details outlined above, should be sufficient for anyone with a workable knowledge and adequate interest in the subject to abstract detailed guidance needed to assemble a suitable pelagic gear.

Fig 42 A 'twin warp method' pelagic pair trawl

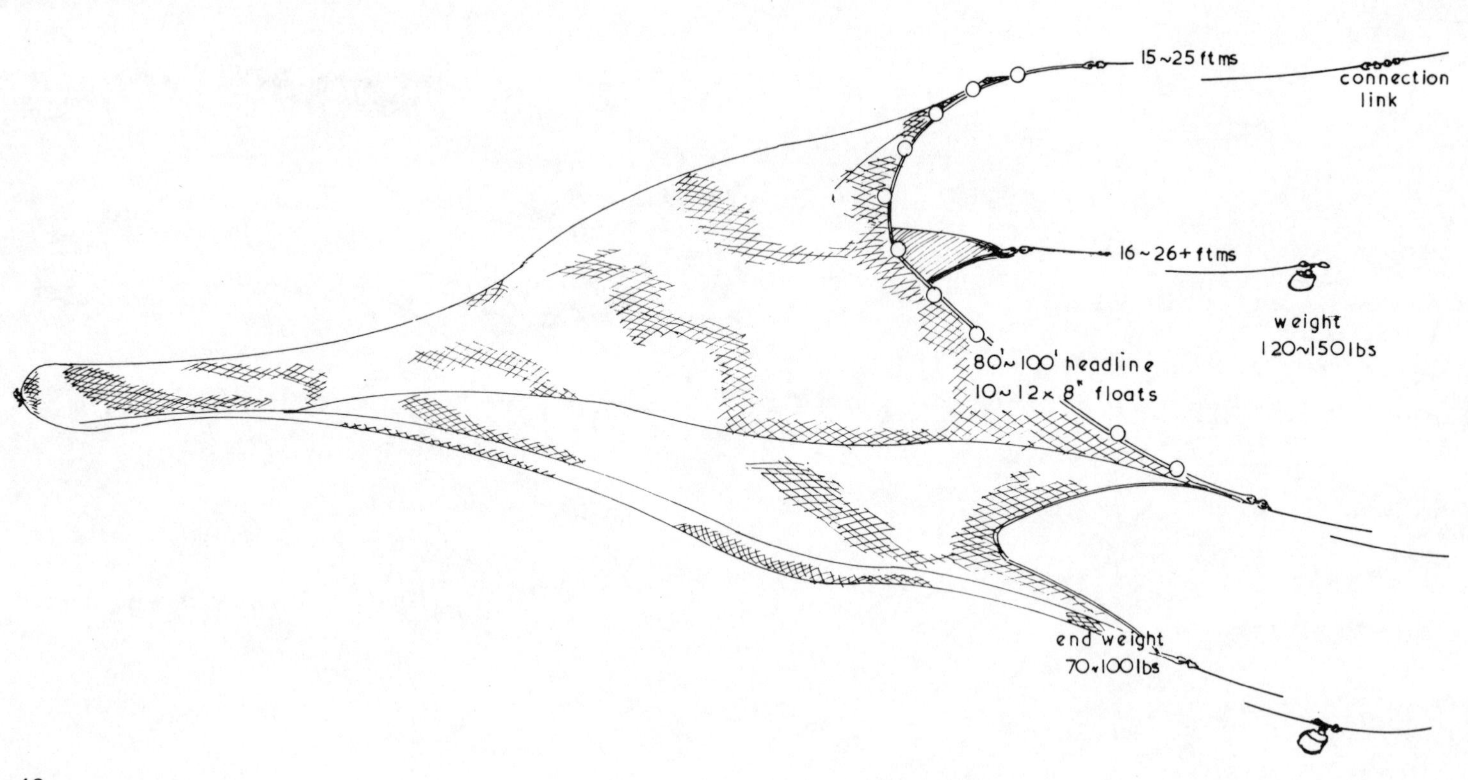

8. Mesh Size and its Selection

Selecting the mesh size of a fishing net or part of a net can be crucial to the satisfactory operation of a fishing assembly. Much, of course, depends on the type of gear and its function. But there are three basic factors common to all. They are as follows:

(1) Size of of the aperture
(2) Dimensions of the material.
(3) Stength/formation of the material.

These elements determine the construction of the meshes that will constitute the netting.

Mesh Size

Mesh size has in the past been the fulcrum on which much of the conservation legislation has been based. And limiting the mesh size in certain areas of the sea to regulated minimums permissible for operation, has been regarded as the major factor for controlling over fishing. The internationally agreed legislation is largely directed at trawling gears which have always been regarded as the most damaging of operations.

It therefore followed from the marine biological information available relating to fish species, their movements, sizes and maturity data, certain minimum meshes could be used in certain areas, and a complicated pattern emerged over the years as international groups renegotiated their agreements.

Quite naturally different countries wanted to protect the interests of their own fishermen, particularly where small species such as sprats and shrimps were involved. Accordingly legislated minimums were small indeed. But whilst some locations may at times be predominantly populated with particular kinds of small fish it is not unlikely for larger fish to pass through those locations.

Further, while conservation methods through mesh size legislation are mainly directed at trawling gears only slight emphasis has been given to types of materials. This being related to the stretch or shrinkage probabilities of particular synthetics. Thus we have 70 mm or 90 mm minimums for local waters and 120 mm or 130 mm for distant waters. In the past I personally have felt that these enforcements have helped to develop ground trawls, which would otherwise have remained heavy and cumbersome for many more years. But nowadays fishermen are more scientific about their trawling gears, and they are selective in the choice of meshes to be used for the various panels.

They do now for instance, consider (1) the size of a mesh in relation to its filterage and load, (2) the suitability of particular twines or cords, their resiliance and the costs involved.

The smallest meshes of a trawl net are naturally those at the tail of the codends where the catch is to be held, and it is there where the minimums are mostly applicable. In this respect it does seem that the present legislation is adequate for normal white fish trawls.

Most examination is probably required on types of gears in relation to their mesh size. For example a purse seine can have an indiscrimate and highly effective percentage catch rate in a shoal of fish, but with precisely regulated meshes it could perhaps be one of the most selective fishing gears in use.

Moving to industrial fishing, the industry does, to a large extent, involve semi-pelagic and pelagic trawl gears, the choice of tail being the governing factor for the species to be caught.

Figure 43 shows a vessel 'aiming' into a shoal exemplifying the mesh size graduation down to the tail of the trawl. Similarly Figure 44 illustrates the specification of a single boat pelagic trawl for

1000/1200 hp trawlers. It is a popular trawl presently in use in the North Sea and other near waters. It presents the severe decrease in mesh size from 80 cm in the wings down to 3 cm in the codends or tail. This type of net is also used from two boats of 350 hp–400 hp (the lower towing power being possible because of reduced drag without otter boards) although as explained in earlier chapters the wings would be more suitable for pair trawling if shaped outwards to follow the 'line of tow' rather than straight.

Dimensions of Materials

1978 does mark the half century since the discovery of nylon, although ropes and twines of this material only appeared in the fishing industry at the end of World War II in the form of surplus parachute cords which were braided up as codends *etc*. Later terylene and polythene appeared on the market.

Of the chemical fibres which were being discovered, certain synthetic polymers proved entirely satisfactory for the fishing industry. Under

Fig 43 One boat pelagic trawl aiming towards shoal

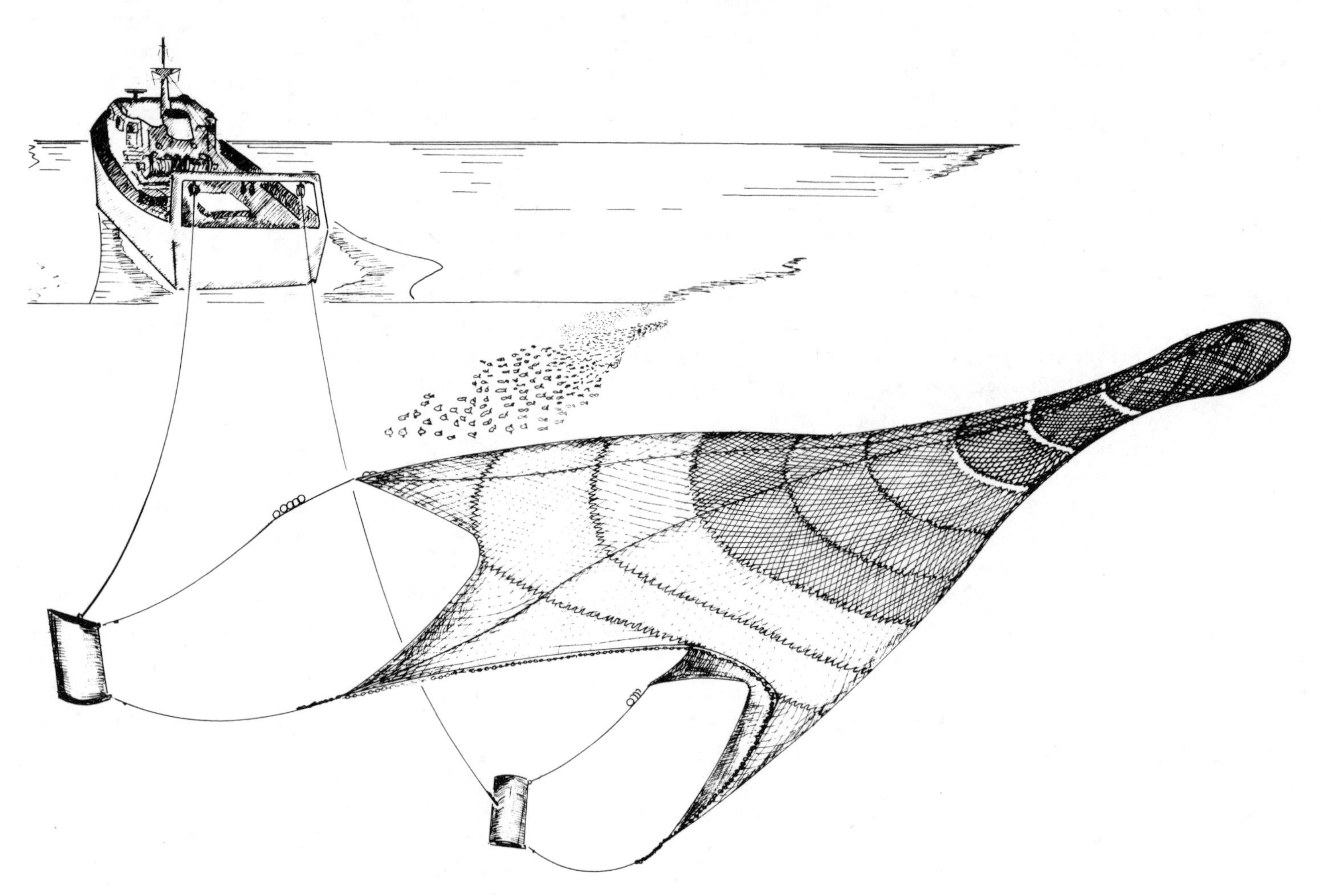

poly-condensation compounds came the polyester fibre terylene, and polyamide materials such as nylon, perlon and niplon *etc*. Whilst the polymeric compounds included in their range polyvinyl alcohol fibres like vinylon and kuralon and perhaps the most successful for the trawling industry those of polyethylene including polythene and courlene.

At this juncture it may be interesting for the reader to know something of the operational differences regarding polyamide type material in contrast to polyethylene makes. Ground trawls would normally be manufactured of polyethylene materials, as being harder wearing and more servicable (codend sections *etc*, may be of nylon or terylene for the strength

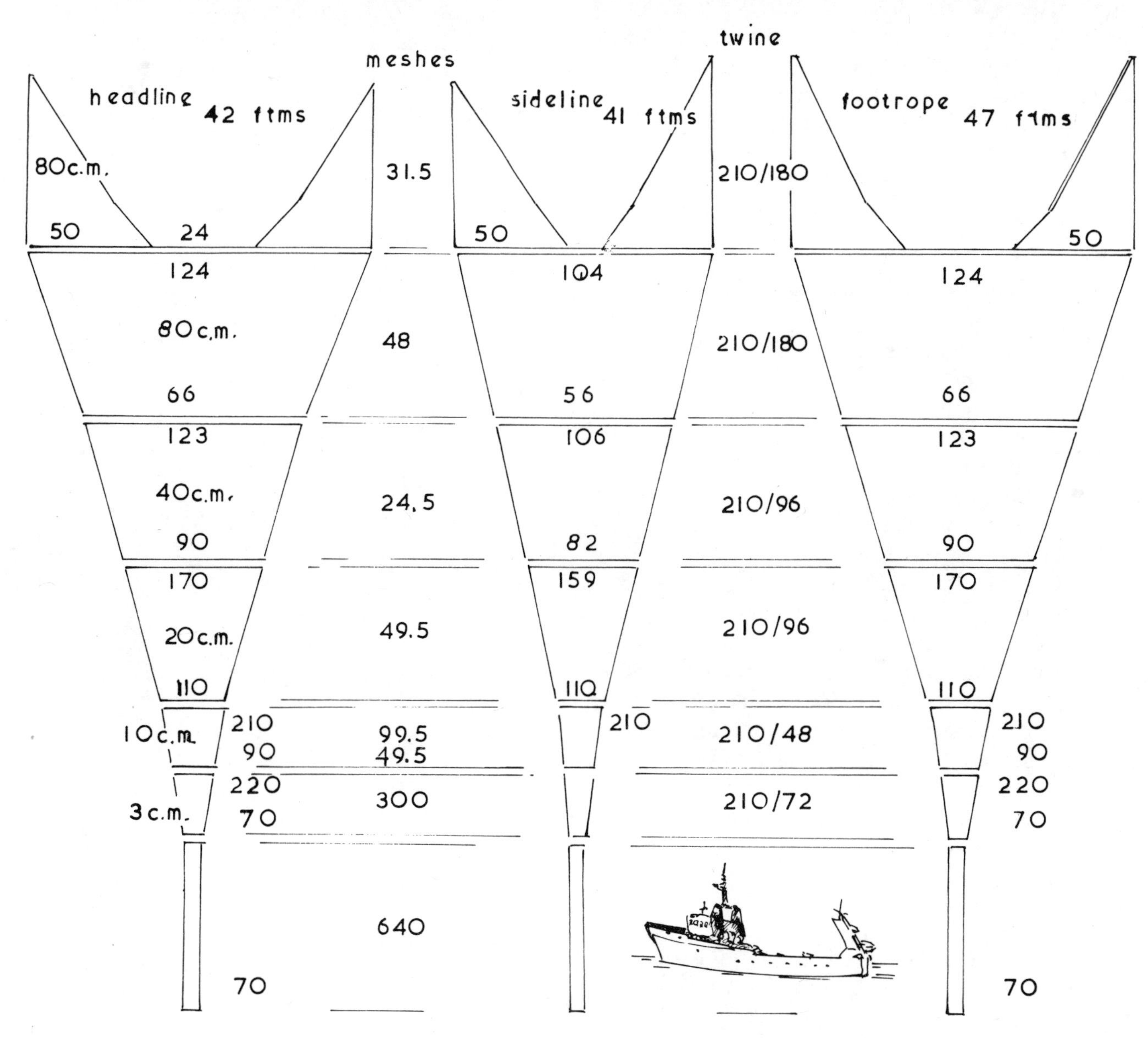

Fig 44 A single boat pelagic trawl for 1000/1200hp vessel

required for lifting the catch out of the water) whereas pelagic trawls would always be specified of polyamide materials such as nylon throughout. Semi-pelagic trawls may probably be constructed of mixed materials, depending on closeness of the fishing contact with the seabed.

When semi-pelagic trawls were first introduced into Britain as wing trawls, in the late 1950's, many were made of nylon although most were of courlene. Because at that time these nets were being fished as ground trawls, those of courlene manufacture were very effective, whilst their nylon equivalents were virtually non-productive. Had the nylon wing trawls been rigged for semi-pelagic fishing however, they would presumably have achieved great success.

Strength and formation of materials.

The tenacity of a twine for making a mesh is obviously a major factor to be considered when making a net, but it is not necessarily the 'all important one'. Resistance to abrasion can be equally important. The main endeavour is always to select the right thickness with the required strength for the particular section of the trawl net.

With mechanisation most trawl net twines were spun and twisted, then laid into ranges determined by their runnages to weight, such as number of yards to the pound or number of metres to the kilo. Later more plaiting machinery appeared around the world and it may be the situation today that the vast majority of pelagic type trawls are now mostly, if not fully, produced of plaited polyamide twines.

In some cases film extruded strips of synthetic materials have replaced fibres as laid twines. Twine of this sort is more susceptible to make the mesh distort or bulge and possibly reduce its normally expected size during operation.

Note: Various pieces of equipment have been evolved for measuring mesh size, inside and outside, and no doubt more sophisticated aparatus is still to come in order to enforce existing legislation.

Fig 45 Pelagic trawling with one or two vessels

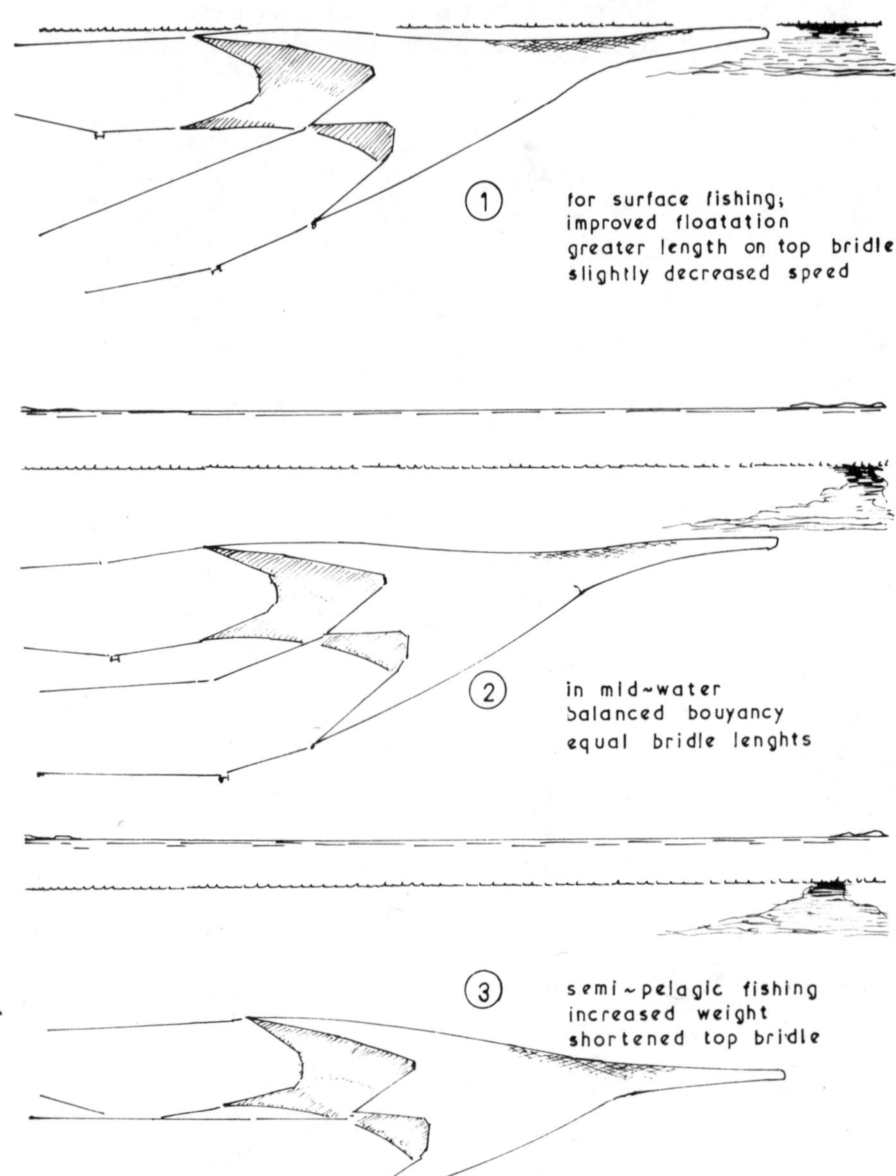

Appendix I

Glossary of fishing gear terms
Glossaire de termes d'engins de pêche
Glosario de términos de aparejos de pesca

Extracted from FAO Catalogue of Fishing Gear Designs (published by Fishing News Books Ltd) with the kind permission of FAO.

	English	Français	Español
A.	**aimed trawling**	*chalutage contrôlé*	arrastre dirigido
	angle of attack (of trawl board)	*angle d'attaque (de panneau de chalut)*	ángulo de ataque (de puertas de arte de arrastre)
B.	**backstrop**	*patte (de panneau de chalut)*	pata de gallo (de las puertas)
	bag, bunt	*poche, sac*	copo del arte
	bagnet	*filet trappe*	red de copo
	bait	*appât*	cebo
	baitings	*(1) diminutions*	(1) reducción de mallas
		(2) petit dos	(2) casarete, cazarete
	ballast	*lest*	lastre
	bar (of mesh)	*patte (côtré de la maille)*	pie (lado de la malla)
	becket	*erse*	estrobo
	belly (of trawlnet)	*ventre (de chalut)*	vientre (de arte de arrastre)
	bellyline	*barrette de ventre*	barreta
	bobbin (of groundrope)	*diabolos, sphères (de bourrelet)*	bolas, diábolos, esféricos (de relinga inferior o de plomos)
	bolchline	*filière*	trencilla
	bonding	*fixation*	fijación
	bosom (of trawlnet)	*carré (de chalut)*	bocana (de arte de arrastre)
	bracket (of trawl board)	*branchon (de panneau de chalut)*	triángulo (de puerta de arte de arrastre)

	English	French	Spanish
	braiding (of netting)	*laçage (de filet)*	trenzado (de la red)
	branchline	*avançon*	brazolada
	breaking load	*résistance à la rupture*	resistencia a la rotura
	buoyancy	*flottabilité*	flotabilidad
	buoylight	*feu ou fanal de bouée*	boya luminosa
	butterfly (of trawl)	*guindineau, triangle (de chalut)*	calón (de red de arrastre)
C.	**chafer (for codend)**	*tablier (pour cul de chalut)*	parpala (del copo)
	codend (of trawlnet)	*cul (de chalut)*	saco, copo (de arte de arrastre)
	codline (of trawlnet)	*raban de cul (de chalut)*	sereta (de arte de arrastre)
	coil (of rope)	*glène (de filin)*	rollo (de cabo)
	coir	*coco*	coco
	combination rope	*filin mixte*	alambrada, cabo mixto
	crowfoot	*patte d'oie*	pie de gallo
	cutting rate (of netting)	*processus de coupe (de filet)*	tipo de corte (de paños de red)
D.	**danleno**	*guindineau*	calón
	depressor	*plongeur*	depresor
	disc roller (for groundrope)	*diabolo plat (pour bourrelet)*	diábolo, rodillo (de relinga de plomos)
	double knot	*double noeud*	nudo doble
E.	**elasticity (of netting yarn)**	*élasticité (de fil pour filet)*	elasticidad (del hilo para red)
	elongation (of netting yarn)	*allongement (de fil pour filet)*	alargamiento (del hilo para red)
	end bracket	*gousset d'extrémité*	pie de gallo
	eye splice	*oeil épissé*	gaza
F.	**fishing lamp**	*lampe de pêche*	lámpara de pesca
	fish pump	*pompe à poisson*	bomba para peces
	flapper (of trawlnet)	*voile, tambour (de chalut)*	trampa (de la red de arrastre)
	float	*flotteur*	flotador
	floatline	*ralingue de flotteurs*	relinga alta, de corchos
	flying mesh or flymesh	*maille folle*	malla libre, angola
	foam plastic	*mousse de plastique*	plástico poroso

	footrope	*bourrelet, ralingue inférieure*	relinga inferior
	front weight	*poids antérieur*	peso anterior
	full mesh (in cutting of netting)	*maille franche (en coupe de filet)*	malla franca (en la reducción de la red)
	funnel	*entonnoir*	embudo
G.	**gear (for fishing)**	*engin (pour la pêche)*	arte de pesca
	G-hook	*croc en G*	gancho en G
	grassrope (of trawl)	*bourrelet de coco (de chalut)*	relinga de bonote (de arte de arrastre)
	groundrope	*bourrelet*	relinga de plomos
H.	**halving back, see splitting strop**		
	hanging (of netting)	*montage, armement (d'un filet)*	armar un arte
	hanging ratio	*taux d'armement*	coeficiente de armadura
	headline	*corde de dos*	relinga de corchos
	heaving bag	*double poche extérieure*	saco de izar
	high opening trawl	*chalut à grande ouverture verticale*	arte de arrastre du mucha abertura vertical
	hoop	*cerceau*	cercar, rodear, cerco
	hydrofoil (trawl board)	*hydrodynamique (panneau de chalut)*	hidrodinámica (puerta de arte de arrastre)
I.	**inflatable float**	*flotteur gonflable*	flotador
J.	**jig**	*faux*	muestra
	joining (of net sections)	*abouture, collage (de pièces de filet)*	unión (de paños de red)
K.	**kelly's eye**	*huit*	ocho
	kite	*plateau élévateur*	elevador
	knot	*noeud*	nudo
	knotless netting, Raschel	*filet sans noeuds, Raschel*	red sin nudos, Raschel
	knotless netting, twisted	*filet sans noeuds, retordu*	red sin nudos, colchada
L.	**lacing**	*transfilage*	pasar una randa, ligadura, atadura
	lastridge (of trawlnet)	*ralingue de côté (de chalut)*	relinga de costado (de arte de arrastre)

	English	French	Spanish
	lay (of rope, etc.)	*commettage (d'un filin, etc.)*	colchado (de un cabo, etc.)
	lazyline	*baîllon, hale-à-bord*	vira-vira
	leadline	*ralingue plombée*	relinga de plomos
	leg (of trawl)	*patte (de chalut)*	pernada
	lengthening piece (of trawlnet)	*rallonge (de chalut)*	antecopo, manga (de arte de arrastre)
	line	*ligne*	línea
	links (of chain)	*maillons (de chaîne)*	eslabones (de cadena)
M.	**mesh**	*maille*	malla
	meshsize (stretched)	*longueur de la maille (étirée)*	longitud de malla (estirada)
	midwater trawl	*chalut pélagique*	arte de arrastre pelágico
	monofilament	*monofilament*	monofilamento
	multifilament	*multifilament*	multifilamento
N.	**net**	*filet*	red
	netting	*nappe de filet, alèze*	paño de red
	netting yarn	*fil pour filet*	hilo para redes
O.	**otter board, see trawl board**		
	otter trawl	*chalut à panneaux*	arte de arrastre de puertas
	overhand knot	*noeud simple*	nudo llano
P.	**pair trawl**	*chalut-boeuf*	arrastre de pareja
	panel (of net)	*face (de filet)*	paño
	patent link	*maillon breveté*	eslabón de patente
	pelagic trawl, see midwater trawl		
	pennant	*rapporteur*	falsa boza
	plaited (netting yarn)	*tressé (fil pour filet)*	trenzado (hilo para redes)
	point (in cutting of netting)	*maille de côté (en coupe de filet)*	malla de costado (corte de paños)
	poke, pork line, see lazyline		
	pole and line (tuna fishing)	*canne (pêche du thon)*	pesca con caña (pesca de atún)

Q.	**quarter point**	*triangle d'aile (au-coin de carré)*	triángulo de banda
	quarter rope	*parpaillot, biribi*	vira-vira, parpallón
R.	**Raschel, see knotless netting**		
	recessed link	*maille à méplats*	eslabón ranurado
	reef knot	*noeud plat*	nudo llano
	rig (of gear)	*gréement (d'engin)*	armazón (del arte)
	roller (for groundrope)	*diabolo (pour bourrelet)*	diábolo, rodillo (para relinga de plomos)
	rope	*filin, cordage*	cabo
	runnage	*longeur par unité de poids*	longitud por unidad de peso
S.	**seam (of net)**	*couture (de filet)*	costura (de red)
	selvedge	*bordure renforcée*	enchace, borde, costura
	semi-pelagic trawl	*chalut semi-pélagique*	arte de arrastre semipelágico
	shackle	*manille*	grillete
	shear-chain	*chaîne d'écartement*	cadena de refuerzo
	sheet bend, or weaver knot	*noeud d'écoute ou de tisserand*	vuelta de escota, nudo de tejedor
	shrinkage (of yarn, etc.)	*retrait au mouillag (de fil, etc.)*	contracción
	side seam	*couture latérale*	costura lateral
	side trawler	*chalutier latéral*	arrastrero por el costado
	sinker	*lest*	plomo
	snood	*empile*	tanza
	spacer disc (for groundrope)	*intermédiare (de bourrelet)*	separador (del burlón)
	splitting strop	*erse de cul*	estrobo de saco
	square (of trawlnet)	*grand dos (de chalut)*	cielo, visera
	staple fibre	*fibre discontinue, schappe*	fibra corta
	sterntrawler	*chalutier arrière*	arrastrero por popa
	strand (of yarns)	*toron (de fil)*	cordón (de hilo)
	strengthening rope	*filin de renfort, ralingue*	cabo de refuerzo, relinga

	strip (of netting)	*bande (de filet) ou nappe*	paño (piezas de redes)
	strop	*erse*	estrobo
	swivel	*émerillon*	grillete giratorio
T.	**take-up (of meshes)**	*recrue (de mailles)*	aumento (de mallas)
	tapering	*diminution*	reducción
	taper ratio	*rapport de diminution*	índice de reducción
	trawl board	*panneau de chalut*	puerta del arte
	trawler	*chalutier*	arrastrero
	trawl gear	*engin de chalutage*	arte de arrastre
	twine	*fil, fil retors*	hilo
	twist factor (of yarn)	*coefficient de torsion (de fil)*	coeficiente de torsión (del hilo)
V.	**vinge trawl, see wing trawl**		
W.	**warp (for trawl)**	*fune (de chalut)*	cable de arrastre
	weaver knot (or sheet bend)	*noeud de tisserand (ou noeud d'écoute)*	nudo de tejedor (vuelta de escota)
	webbing, see netting		
	wing (of trawlnet)	*aile (de chalut)*	ala, bandas, pernadas
	wingtip	*pointe d'aile*	extremo de la banda
	wing trawl	*chalut à grande ouverture verticale*	red a gran abertura vertical
	wire rope	*filin d'acier*	cable, cable de acero
Y.	**yarn, see netting yarn**		
Z.	**zipper line**	*ligne de trasfilage*	matafión, cabos para dividir los cercos de jareta
	zipper ring	*anneau pour transfilage*	anillas de los cabos de división del gran cerco de jareta

Appendix II

CONVERSION TABLE

(Approximate Figures–Ed)

Inches	*Millimetres*	*Feet*	*Metres*	*Fathoms*
(½) 0.5				
(¾) 0.75				
*1	25.4	0.08	0.02	0.01
2	50	0.16	0.05	0.03
3	76	0.25	0.07	0.04
4	101	0.33	0.10	0.06
5	125	0.42	0.12	0.07
6	152	0.50	0.15	0.08
7	177	0.58	0.17	0.10
8	203	0.66	0.20	0.11
9	229	0.75	0.23	0.13
10	254	0.83	0.25	0.14
11	280	0.92	0.28	0.15
12	305	*1.00	0.30	0.17
18		1.50	0.45	0.25
24		2.00	0.60	0.33
36		3.00	*0.91	0.50
39.4		3.35	*1.00	0.56
54		4.50	1.37	0.75
72		6.00	1.82	*1.00
144		12.00	3.66	2.00

* 0.9144 metre – 1 yard (3 feet)
* 1.8288 metre – 1 fathom
* 1.094 yard – 1 metre
* 25.4 millimetre – 1 inch

Other books published by Fishing News Books Limited Farnham, Surrey, England

Free catalogue available on request

A living from lobsters
Aquaculture practices in Taiwan
Better angling with simple science
British freshwater fishes
Coastal aquaculture in the Indo-Pacific region
Commercial fishing methods
Control of fish quality
Culture of bivalve molluscs
Eel capture, culture, processing and marketing
Eel culture
European inland water fish: a multilingual catalogue
FAO catalogue of fishing gear designs
FAO catalogue of small scale fishing gear
FAO investigates ferro-cement fishing craft
Farming the edge of the sea
Fish and shellfish farming in coastal waters
Fish catching methods of the world
Fish farming international No 2
Fish inspection and quality control
Fisheries oceanography
Fishery products
Fishing boats of the world 1
Fishing boats of the world 2
Fishing boats of the world 3
Fishing ports and markets
Fishing with electricity
Fishing with light
Freezing and irradiation of fish
Handbook of trout and salmon diseases
Handy medical guide for seafarers
How to make and set nets
Inshore fishing: its skills, risks, rewards
International regulation of marine fisheries: a study of regional fisheries organizations
Marine pollution and sea life
Mechanization of small fishing craft
Mending of fishing nets
Modern deep sea trawling gear
Modern fishing gear of the world 1
Modern fishing gear of the world 2
Modern fishing gear of the world 3
Modern inshore fishing gear
More Scottish fishing craft and their work
Multilingual dictionary of fish and fish products
Navigation primer for fishermen
Netting materials for fishing gear
Pair trawling and pair seining – the technology of two boat fishing
Planning of aquaculture development – an introductory guide
Power transmission and automation for ships and submersibles
Refrigeration on fishing vessels
Salmon and trout farming in Norway
Salmon fisheries of Scotland
Seafood fishing for amateur and professional
Ships' gear 66
Sonar in fisheries: a forward look
Stability and trim of fishing vessels
Testing the freshness of frozen fish
Textbook of fish culture; breeding and cultivation of fish
The fertile sea
The fish resources of the ocean
The fishing cadet's handbook
The lemon sole
The marketing of shellfish
The seine net: its origin, evolution and use
The stern trawler
The stocks of whales
Training fishermen at sea
Trawlermen's handbook
Tuna: distribution and migration
Underwater observation using sonar